B.R. STANDARD LOCOMOTIVES TO SCALE

in this series by Ian Beattie

SOUTHERN LOCOMOTIVES TO SCALE

L.M.S.R. LOCOMOTIVES TO SCALE

B.R. STANDARD LOCOMOTIVES TO SCALE

G.W.R. LOCOMOTIVES TO SCALE

L.N.E.R. LOCOMOTIVES TO SCALE

B.R. STANDARD LOCOMOTIVES

TO SCALE

IAN BEATTIE

D. BRADFORD BARTON LTD

MAJOR DIMENSIONS

CLASS	NUMBER OF CYLINDERS	OVERALL LENGTH	OVERALL WIDTH	ENGINE WHEELBASE	BOILER PITCH	LEADING WHEELS (DIAMETER)	COUPLED WHEELS (DIAMETER)	TRAILING WHEELS (DIAMETER)	TENDER WHEELS (DIAMETER)
BRITANNIA 4-6-2	2	68′ 9″	9′ 0″	35′ 9″	9′ 4″	3′ 0″	6′ 2″	3′ 3½″	3′ 3½″
CLAN 4-6-2	2	68′ 9″	9′ 0″	35′ 9″	9′ 2″	3′ 0″	6′ 2″	3′ 3½″	3′ 3½″
5 4-6-0	2	62′ 7″	8′ 10⅛″	27′ 3″	8′ 9″	3′ 0″	6′ 2″	—	3′ 3½″
4 4-6-0	2	60′ 0″	8′ 7⅜″	26′ 9″	8′ 9″	3′ 0″	5′ 8″	—	3′ 3½″
4 2-6-4T	2	44′ 9⅞″	8′ 10⅛″	36′ 10″	8′ 9″	3′ 0″	5′ 8″	3′ 0″	—
4 2-6-0	2	55′ 10½″	8′ 7⅜″	24′ 1″	8′ 11″	3′ 0″	5′ 3″	—	3′ 3½″
3 2-6-2T	2	40′ 10½″	8′ 6⅝″	32′ 7″	8′ 7½″	3′ 0″	5′ 3″	3′ 0″	—
2 2-6-0	2	53′ 2½″	8′ 6″	22′ 3″	8′ 3″	3′ 0″	5′ 0″	—	3′ 3½″
2 2-6-2T	2	38′ 9½″	8′ 7¾″	30′ 3″	8′ 3″	3′ 0″	5′ 0″	3′ 0″	—
3 2-6-0	2	55′ 11¼″	8′ 6⅝″	24′ 1″	8′ 7½″	3′ 0″	5′ 3″	—	3′ 3½″
8 4-6-2	3	70′ 0″	9′ 0″	36′ 9″	9′ 4″	3′ 0″	6′ 2″	3′ 3½″	3′ 3½″
9 2-10-0	2	66′ 2″	8′ 10⅛″	30′ 2″	9′ 6½″	3′ 0″	5′ 0″	—	3′ 3½″
9 F-C 2-10-0	2	66′ 2″	8′ 11″	30′ 2″	9′ 11¾″	3′ 0″	5′ 0″	—	3′ 3½″

8188/3BL ISBN 0 85153 390 6
printed in Great Britain by Lovell Baines Print Ltd., Newbury, Berkshire and bound by Robert Hartnoll Ltd., Bodmin, for the publishers
D. BRADFORD BARTON LTD TRETHELLAN HOUSE TRURO CORNWALL ENGLAND

CONTENTS

drawings to 4mm to 1 foot scale, except for those showing details

LIVERY AND INSIGNIA

In the years immediately following nationalisation, before the appearance of the Standard classes, several experimental liveries were applied – in particular to express passenger locomotives – in the quest for a definitive B.R. colour scheme. Eventually the following liveries were adopted for the steam fleet, including of course the Standard locomotives.

EXPRESS PASSENGER: Ex-Great Western Brunswick Green on boiler, cab, tender side, rear and top plates, and vertical running-plate surfaces. The locomotives within this category were fully lined-out, again using the Great Western scheme of black sandwiched by orange. Cylinder covers were painted black with twin vertical red lines on cladding bands.

PASSENGER AND MIXED TRAFFIC: Black overall, with L.N.W.R.-derived lining in grey (outer), cream and red (inner) lines, with boiler bands lined in red only. Unlike the express passenger locomotives, lining was also applied to running-plate vertical surfaces. Cylinder covers lined as above. In several instances – principally on the Western Region – passenger-working locomotives were finished in unlined green.

FREIGHT, SHUNTING AND GENERAL-PURPOSE: Black overall, without lining.

Common to all categories of steam locomotive:

Black: Smokebox, chimney, cylinders, locomotive and tender frames, bogie and pony-truck frames, running-plate top surfaces, axleboxes and cab roof. Red: Buffer beams and stocks.

Bright Metal: Wheels, tyres, all motion linkage and rodding. Very often external copper pipework was burnished.

NAMEPLATES: On those Standard locomotives so identified, nameplates comprised plain sans serif script on a black background with raised edging. There were a few instances of red backgrounds. Exception: Class 9 2-10-0 No.92220 'Evening Star' in G.W.R.-style 'bold serif' script.

NUMBERS: On all locomotives: plate on smokebox door. Tender locomotives: on cab sides, painted or decals. Tank locomotives: on bunker sides, painted or decals. The smokebox number plate comprised raised numbers in sans serif script, picked out in white on a black background without edging. The cab/bunker-side numerals were likewise white, sans serif.

ADDITIONAL PLATES: All locomotives carried M.P.D. identification plates, affixed to the smokebox door. These plates, showing the identification number of the Motive Power Depot to which the locomotive was assigned, were oval, with raised white sans serif script on a black background. The B.R. Standard classes additionally carried manufacturing plates, in oval form, giving year of construction and erecting shop identity, on the front frames above the buffer beam running plates. White sans serif script on black background.
Tenders wore two separate plates showing tender type and serial number, attached to the tank backplates. Again white, sans serif script on black background.
Although all twelve Standard classes (bar the Class 9 2-10-0 and the unique Class 8 Pacific) were designated mixed-traffic, liveries appearing on the classes varied according to status, for example, lined green where the proportion of premium passenger work undertaken was relatively high, or where freight work preponderated, etc.

INSIGNIA: Three successive modes of insignia were utilised on the steam locomotives operated by British Railways. The earliest and most transient form was simply the words 'British Railways' rendered in white sans serif script on tender sides (tank sides in case of tank locomotives).
Discounting this rapidly-superseded style, there were two types of emblems:
Lion surmounting a spoked wheel bisected by horizontal bar carrying the 'British Railways' legend. Applied 1949-1957.
Lion clutching a spoked wheel atop a crown, the whole within a circle, flanked by rectangles bearing the legend. In use 1957 until beyond the steam era.
The two emblems co-existed up to the end of steam on British Railways, the later variety being applied to locomotives built or repainted from the late 1950s onwards.

LIVERIES:
Class 7 Britannia 4-6-2; fully lined green.
Class 6 Clan 4-6-2; fully lined green.
Class 5 4-6-0: Lined black; several examples outfitted in fully lined green.
Class 4 4-6-0: Black, lined. A number turned out in lined green.
Class 4 2-6-4T: Lined black.
Class 4 2-6-0: Black, lined.
Class 3 2-6-2T: Lined black. Several were finished in both lined and unlined green – principally those on the Western Region.
Class 2 2-6-0: Lined black, but with cylinders and boiler bands plain. The first of the class, No.78000, was rendered in unlined green.
Class 2 2-6-2T: Black, lined, cylinders and boiler bands plain.
Class 3 2-6-0: Lined black.
Class 8 4-6-2: Fully lined green.
Class 9 2-10-0: Black, unlined. Exception: No.92220 'Evening Star', outshopped in express passenger livery plus copper chimney and buffed copper pipework.
Class 9 Franco-Crosti boilered 2-10-0: Unlined black.

This volume considers the final designs of steam locomotives built for use in this country, the twelve Standard classes produced by British Railways during the 1950s.

Whereas the underlying motive for this book serves to illustrate each locomotive class by way of detailed drawings, photographs and text as an aid to the railway modeller, the general railway enthusiast is by no means neglected; the line drawings depart from the conventional 'blueprint' approach, with the mass of lines and figures that characterise the latter, and are instead produced in a form to appeal to the less technically-minded reader. Nonetheless strict accuracy and attention to detail have been adhered to throughout.

The political conception of a state-owned railway monopoly originated in Britain during the nineteenth century, but was not put into practice until 1947 by passage through Parliament of the requisite Government-inspired Transport Bill. Thus on 1 January 1948 not only the four railway Groups but also road haulage, canals and docks were placed under the control of a government-sponsored organisation, the British Transport Commission (B.T.C.). The Groups had fought valiantly to retain their autonomy but their case was undermined by a number of factors, including the ravages of wartime damage that required government subsidy for repair, lack of staff due to war depletion and subsequent dissatisfaction with working conditions involved, undervalued reimbursement for wartime haulage services, unreasonably low freight charges compulsorily forced on the companies by the immediate pre-war Government and still in force thereafter; the railway Unions, moreover, favoured nationalisation.

In common with all too many enterprises run by committees with the resultant lack of decision, communication or positive leadership, British Railways suffered an uneasy and hesitant initial period: persistent governmental policy redirections and financial waywardness have reaped the inexorable consequences.

The undertaking was split into Regions approximating to the spheres of operation enjoyed by the Groups. Each Region was operated by its own management under the direction of the Railway Executive, a body responsible to the B.T.C., a situation that rapidly deteriorated into dichotomy between regional managers and the Executive since the former tended to continue the ingrained practices of the Groups—none more so than the Western Region, inheritor of the 113-year G.W.R. with its deeply-entrenched traditions and idiosyncracies. Following abolition of the Executive in 1953, full control of B.R. passed to B.T.C. (subsequently the British Railways Board) which in turn was relieved of the other forms of land transport for which it had originally been promoted.

The inception of the B.R. Standard locomotive classes is subject to almost the same level of controversy that has attended the act of nationalisation of the railways. Both the G.W.R. and L.M.S. had standardised fleets (the L.N.E.R. activated standardisation too late for effectiveness) and the appearance of another set of standards added to the burden on locomotive crews as well as shed staffs. In addition, the locomotive erecting programmes in progress immediately prior to nationalisation could not be economically truncated, even though in strict business terms this was a prerequisite for substitution of a new standard.

At the time permission was granted for enactment of the B.R. Standard building schedule, no overall motive power policy had been formulated; the eventual change to diesel and electric traction had been foreshadowed by American and European experiences, and by the turn of the 1940s the sharply rising cost of coal and labour-intensive nature of steam operation had pointed the inevitable way for the future.

A comprehensive series of inter-Regional locomotive exchanges was conducted in 1948 in order to assess individual features in the formulation of an overall Standard adaptable for working anywhere in the U.K. The inherent staff shortage situation necessitated a built-in low maintenance factor which demanded an essentially simple and robust locomotive with emphasis on accessibility to moving parts and maximum use of labour-saving devices. The complication of three or more cylinders was ruled out; temperamental or expensive methods in the production of steam were similarly deleted and a standardised, left-hand drive cab layout was adopted, allowing the driver to control his charge whilst seated. Economic operation of the locomotives themselves was a prime consideration, thus free-steaming boiler and smoothly-regulated power application qualities were paramount, with minimal friction in moving parts via roller bearings or generous plain bearing surfaces a necessary adjunct.

Initially planned for ten classes, the B.R. Standard locomotives were conceived by a panel presided over by Robert A. Riddles, formerly a senior engineer on the L.M.S. and creator of the wartime Austerities. The preponderance of ex-L.M.S. personnel on the design team was not unexpected for the L.M.S. was much the largest of the four Groups, and the resultant locomotives bore a marked L.M.S. orientation. The continuance of pure L.M.S. products in build at nationalisation had been mooted for adoption as B.R. Standards, and would have proved less expensive in the final analysis, yet the disparate loyalties of operational staff hailing from the other Groups mitigated against this course of action.

Detail design work was allotted to the four major locomotive centres inherited by B.R. – Swindon, Derby, Doncaster and Brighton – each being responsible for one or more aspects of the overall specification and, moreover, each class was assigned a central office from the four where principal construction drawings were prepared and collated.

Riddles retired in 1954, before the last of the classes for which he was ultimately accountable was introduced. Although one of a committee, his name alone will always be associated with the 999 B.R. Standard steam locomotives built between 1951 and 1960, all having common elements in their generally well-proportioned appearance – the traditional elegance of British steam power had been foresworn by unusually high-set running plates and a mass of external appendages in the cause of accessibility. That the fleet was in service for so short a time was not merely a waste of manpower, effort and materials but more especially an indictment of the wavering and inharmonious managerial attitudes unfortunately rife in the early history of British Railways.

The unveiling in 1951 at the Festival of Britain exhibition of the first B.R. Standard locomotive prompted a great deal of interest among railway aficionados and general public alike, for it represented the design ideals of the then youthful nationalised railway organisation and indicated the future trend of steam power in Britain.

There was nothing revolutionary in either general appearance or specific detailing, rather a departure from tradition in being an amalgam of familiar fittings and fixtures simplified and resited for minimum maintenance plus maximum accessibility. The use of two cylinders only was a surprise to those long acquainted with three- and four-cylinder Pacifics and inevitably resulted in 'hammer-blow', the deleterious effect of which was alleviated by a new type of rail then being laid, of much enhanced durability than hitherto.

Before construction of the 4-6-2s commenced – the Britannia class title, though unofficial, was derived from the name given to the first locomotive in the series – a mock-up of the projected cab layout was studied by interested parties, not least of whom were locomotive crews and senior shed staff, the arrangement of which was finalised on the comments made thereby. In typical trans-Atlantic fashion but relatively novel to British practice, the cab was attached directly to the firebox rather than the main frames; however the lightweight structure gave rise to high noise levels when running, and in addition severe draughts experienced on the footplate justified eventual fitment of flexible screens between cab and tender (the latter equipped with a spectacle plate matching the cab outline). The requisite qualities of high, easily-regulated steam production and free running via roller bearings were amply borne out in practice and on the Eastern Region where thirteen of the initial batch of fifteen Britannias were consigned, express passenger services were improved out of all recognition – indeed the universally successful Pacifics enjoyed their highest esteem on that Region.

The equable axle rating allowed the versatile Pacifics to handle traffic on routes over which Group-designed locomotives of that wheel arrangement could never have been permitted; in time the B.R. 4-6-2s ventured on to main lines throughout the country, though their acceptance varied from Region to Region, the most reluctant being the G.W.R.-orientated Western, where crews much preferred their Castles and Kings; more importantly, the left-hand driving position was a major drawback on a system intended for right-hand control.

Built at Crewe and numbered in the 70000 sequence, the class was completed in 1954 and all 55 members were individually named; uniquely for a single class, the names were selected on a variety of themes. Tenders of types B.R. 1A, 1D and 1E were used (1Ds were equipped with steam-powered coal pushers) according to allocation. Teething troubles were not unexpected on what was effectively a brand new, untried design and whilst the basic format of the locomotive was quickly proven, a series of minor initial problems occurred, including coupling rods fracturing on sudden loss of adhesion and pistons and cylinders distorted or smashed by excess water in the steam supply. The most potentially hazardous defects concerned coupled wheels that skewed out of position on the axles and loose crosshead slide bars caused by inaccessible holding bolts. Following an accident on Western Region in which forward visibility had been a contributory factor, smoke deflector handrails were removed from W.R. Britannias and perfunctory handholds substituted; new locomotives were subsequently fitted during construction with deflectors minus handrails.

A curious misadventure befell a Britannia-hauled Eastern Region express at Ilford in August 1957: without warning the locomotive parted company with its tender and train whilst at speed. The train was automatically brought to a stop by the rupture of vacuum brake piping; however, as the locomotive braking system relied on connection with the tender, the runaway engine could not be braked and was only brought to a standstill by rapid action on the part of the driver in putting reverse gear into effect. The culprit: a sheared drawbar coupling pin, thereafter modified to avert similar occurrences.

Although categorised as mixed-traffic locomotives, the Class 7 4-6-2s concentrated primarily on express passenger work and throughout their brief career proved consistently fast, efficient and economical in operation. Scrapping commenced in 1965 and by the end of 1967 all bar one had disappeared into oblivion. The exception was No.70013 'Oliver Cromwell', which took part in the final B.R. steam working in August 1968 and thus bears the dubious distinction of being the last steam express passenger haulier in regular B.R. revenue-earning service. Both No.70013 and the prototype, No.70000 'Britannia' are extant in full working order.

No. 70036 'Boadicea' simmers gently at Stratford shed, Eastern Region in June 1959. Plainly evident is the injector pipework beneath and forward of the cab and the chime whistle on the smokebox flank. *(F. Hornby)*

'BRITANNIA' CLASS 4-6-2, 1951

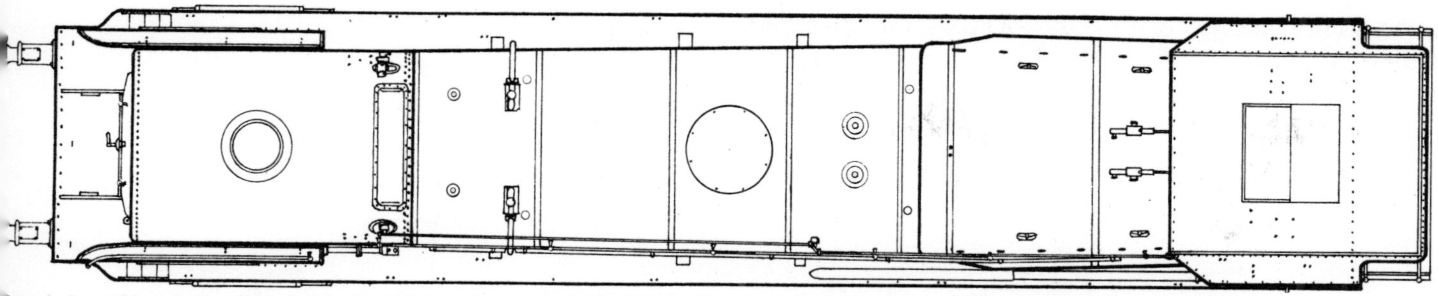

or clarity, all handrails, pipework and rodding on the upper half of the boiler, plus all pipework forward of the cab, have been omitted

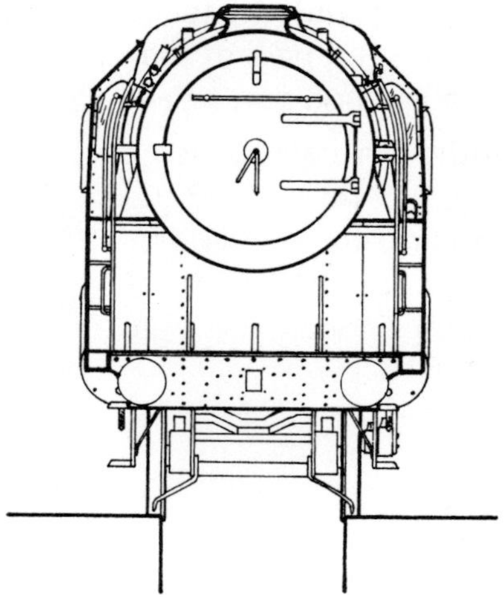

1

'BRITANNIA' CLASS 4-6-2, 1951

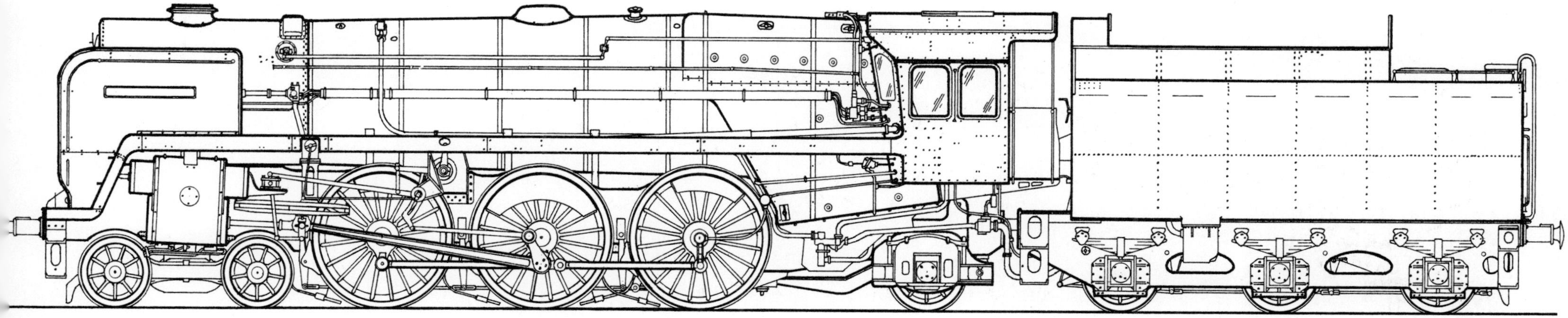

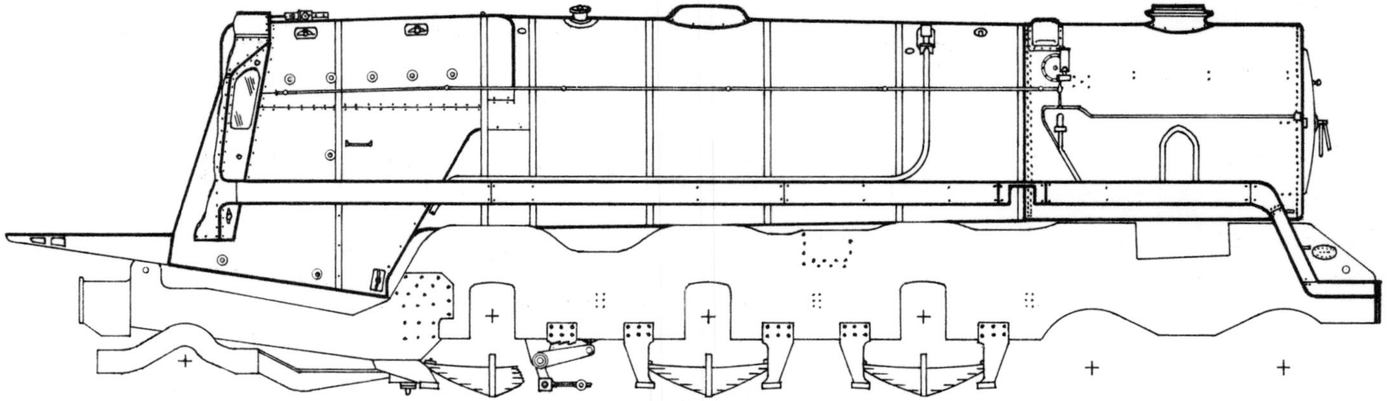

FRAME AND RIGHT-HAND EXTERNAL DETAILS. CAB CUT AWAY TO REVEAL FIREBOX BACKPLATE

SMOKE DEFLECTORS WITH HANDHOLDS
Standard Style
L.M.S.-type handholds, backed by semi-spherical cups
(see front view of plate)
Small horizontal grabrail also fitted

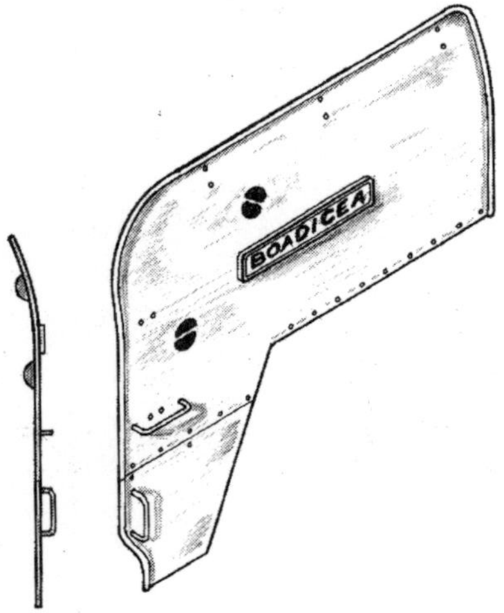

Western Region modification of original deflectors;
handrails removed and handholds substituted, punched
through plating with strengthening rims added

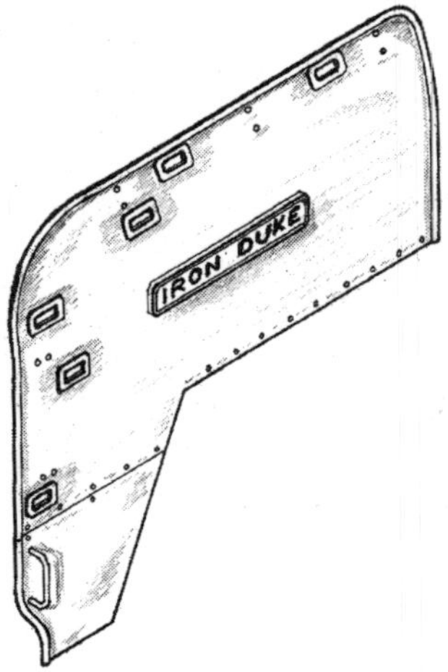

'BRITANNIA' CLASS 4-6-2, 1951

'CLAN' CLASS 4-6-2, 1951

At first glance identical in appearance to the Britannia, the Clan class was Riddles' response to a demand for a 4-6-2 of close design affinity to the Class 7 but with lighter axle weight and intended for working major trains on the Highland line in Scotland. Smaller-diameter cylinders and a smaller boiler with taller boiler mountings comprised the principal diversions from the Class 7 specification, resulting in a Class 6 power rating, and remaining parts being interchangeable between the two Pacifics.

The initial batch of ten engines was outshopped from Crewe in 1951-52 and an order was placed for assembly of a further fifteen. The order was never ratified however, since the levels of performance proved dismal in relation to the locomotives' mechanical promise, and although permitted a wider route availability than the contemporaneous Standard Class 5 4-6-0s were incapable of matching the latter's overall competence. Their usefulness limited and their duplication no longer necessary or desirable, the composition of the class terminated with the original ten engines, which were individually named for Scots clans and numbered in the 72000 series.

The need for a locomotive of this genre was debatable in the first instance. There was a multiplicity of comparable ex-L.M.S. three-cylinder engines in existence that performed similar work allocated to the Clans with acknowledged prowess, and the peculiarities of the Highland line where steep gradients abounded favoured the use of locomotives of higher adhesion in the B.R. Class 5 and Class 7 mould. The poor characteristics in practical operation of the Clans, stemming from difficulties in regaining steam pressure en route if allowed to fall under par, allied to unduly high coal consumption in normal production of steam, led to demotion from prestigious Scots expresses onto routine duties over the Glasgow-Liverpool-Birmingham tracks. No.72009 'Clan Stewart' underwent haulage tests on Eastern Region, very much the preserve of Britannias, on conclusion of which it was smartly conducted back to the Scottish Region . . .

Type B.R.1A tenders were attached to all ten Clans and in some instances conventional whistles replaced original-equipment chime whistles.

This, the least proficient of the twelve B.R. Standard classes, was committed to the scrapyards between 1962 and 1966, its passing unmourned.

Fully prepared for a day's duty, No.72005 'Clan MacGregor' awaits its crew at Carlisle Kingmoor shed, L.M. Region in June 1964. Note the speedometer take-off point on the rear coupling-rod crank. *(F. Hornby)*

'CLAN' CLASS 4-6-2, 1951

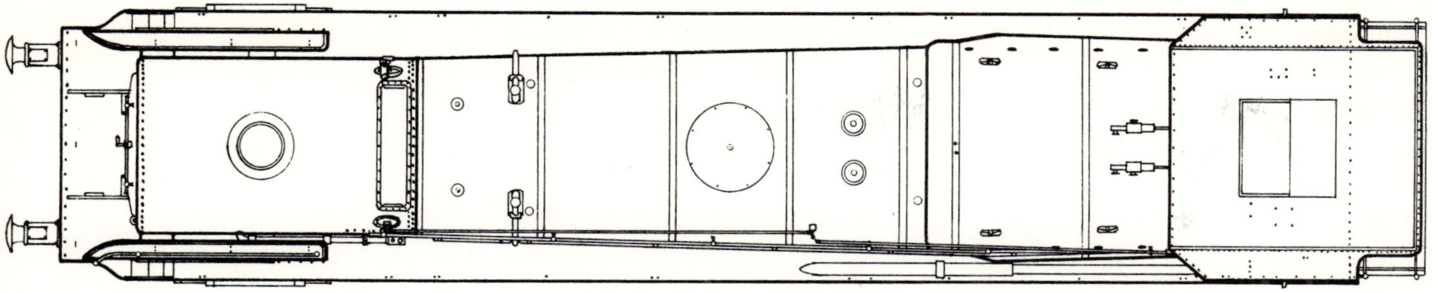

For clarity, handrails on the 'upper' boiler surface, together with all pipework forward of the cab, have been omitted.

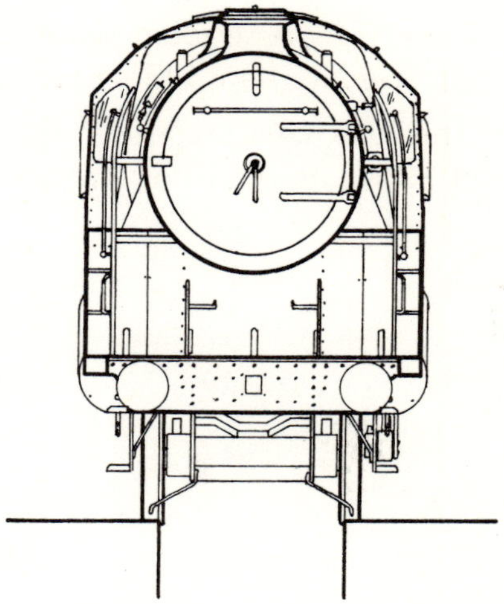

'CLAN' CLASS 4-6-2, 1951

15

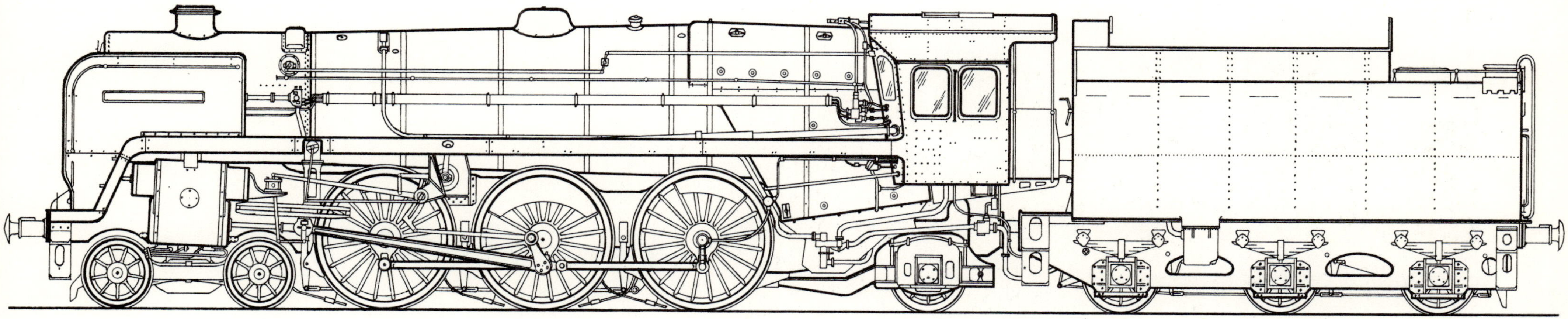

The B.R. equivalent to the G.W.R. Hall and L.M.S. Black Five, and resembling the latter in appearance, the Class 5 4-6-0 was a highly-versatile mixed-traffic haulier incorporating wheels, cylinders (linered down) and valve motion shared with the Pacifics. It also had a boiler descended directly from the L.M.S. Standard as applied to Black Fives and Class 8 2-8-0s, with minimal modification. The latter was evidenced outwardly by external regulator rodding plus the Southern Railway-inspired feed water clack valves common to all B.R. Standard designs.

In view of its close relationship to Stanier's mixed-traffic masterpiece, the necessity for reappraisal of the L.M.S. formula was challenged in some quarters – the traffic situation that prompted building of 800+ Black Fives, a considerable proportion of which were completed under B.R. auspices to within a few months of announcement of the replacement Class 5, was as valid in 1951 as it had been in Stanier's era – however the establishment of a common set of standards in accord with B.R. operational policy pre-empted the continuation of the L.M.S. design.

That the B.R. Class 5s were a success is vindicated by the numbers outshopped: 172 between 1951 and 1957, all emanating from the ex-L.M.S. works at Derby. Distributed widely throughout Britain the engines proved extraordinarily reliable, in good fettle achieved performances more appropriate to vigorous 4-6-2s, and were among the most popular of classes with those most discriminating of railway personnel, the locomotive crews. The cab design, with its angled front window panels and side sheets that sloped gently inwards from a point just below the windows, was identical to the Pacifics and while the noise levels proved less of a nuisance on the Class 5s, the presence of swirling draughts was largely assuaged by fitment of flexible screens, together with reinstatement of a short fall-plate (a steel plate hinged about the trailing edge of the footplate, bridging the gap between engine and tender) that had been deemed unwarranted at the planning stage. Chime whistles were fitted to early members of the class, eventually replaced by straightforward whistles. The majority of 4-6-0s carried their whistles in an unusual location, upright on the smokebox abaft the chimney, though a number of early engines were provided with replacement whistles in the conventional position atop the firebox.

Owing to the broad variation in Regional allocations the Class 5s were equipped with five types of tender, differing in coal and water storage capacities: types B.R.1, 1A, 1B, 1C and 1F. Those 4-6-0s assigned to Southern Region, where water trough facilities were non-existent, were given the ultra-capacious B.R.1F containing 5625-gallon water tanks. Numbered in the 73000 sequence and generally found in lined black livery in accordance with their mixed-traffic status – notwithstanding which several appeared in fully-lined express passenger green, since such exalted work was within their capability and frequently exploited by the running departments – the Class 5s were comparatively heavy locomotives and thus limited to favourable routes. Twenty locomotives bore individual names, all on Southern Region, the nameplates being transferred from scrapped King Arthur Class 4-6-0s and mounted on the running plate valances.

Following experience with the unique Class 8 Pacific, Caprotti valve gear was installed on thirty Class 5s outshopped in 1956. The rotary nature of this means of valve control called for finer manufacturing tolerances and a consequently more generous degree of maintenance than the simple, time-proven Walschaert system, nevertheless the advantageous increase in tractive power displayed by the engines so equipped offset the extra expense involved in installation and upkeep. However the dieselisation programme brought a halt to development in this promising direction.

Progressive retirement of the 4-6-0s was instituted in 1964, the last remaining members of the class being phased out during the death throes of steam in 1968. Three representatives were salvaged for restoration, in commemoration of this meritorious design.

The nicely proportioned lines of the Class 5 4-6-0 are shown to advantage by No. 73082 'Camelot' at Nine Elms shed, Southern Region, in September 1959. (F. Hornby)

CLASS 5 4-6-0, 1951

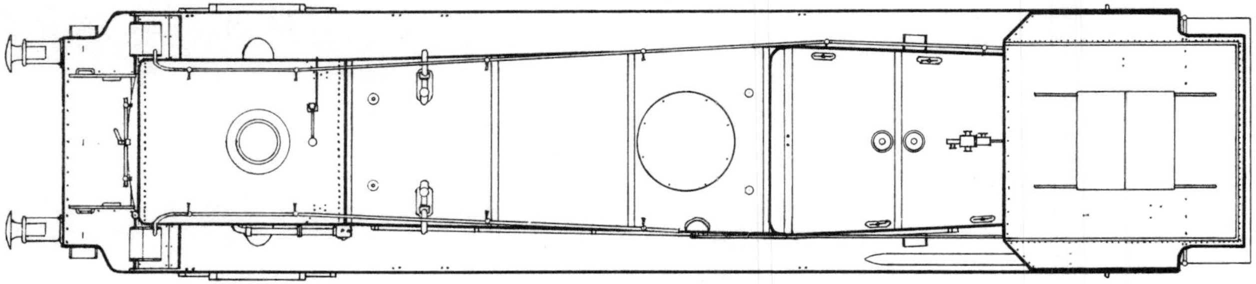

©COPYRIGHT

18 CLASS 5 4-6-0, 1951

CLASS 5 4-6-0, 1951

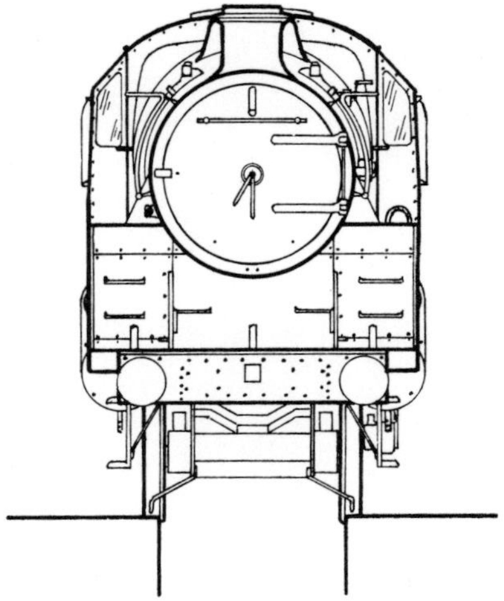

FRAME DETAIL, FIREBOX/ASHPAN DESIGN

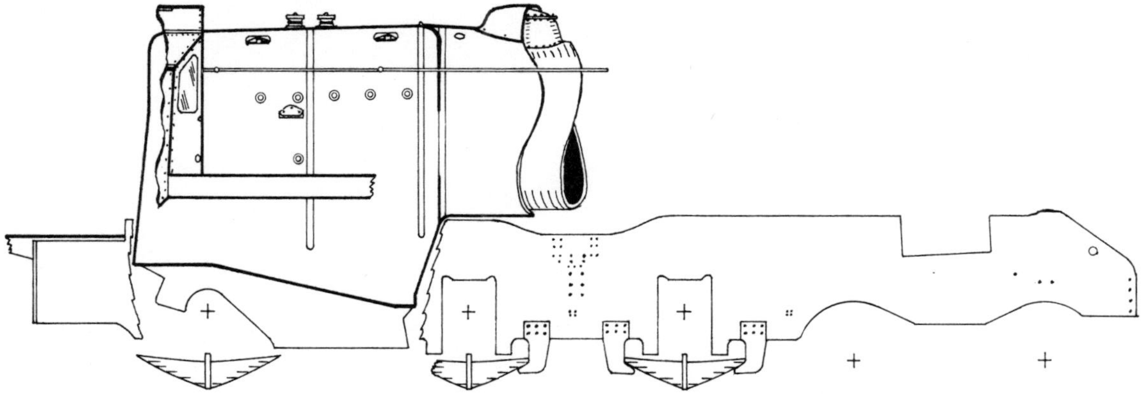

Very similar in appearance and character to the Class 5 4-6-0s, though as their power rating suggests lighter and smaller, the Class 4 versions were developed for use over lines barred to their heavier contemporaries. A comprehensive roster of pre-Nationalisation motive power in the medium-weight category justified a lesser population of B.R. Class 4s relative to the sizable number of Class 5s and coverage was not so widespread, being based chiefly in the Midlands and Southern half of England. Here the locomotives' excellent manoeuvrability was put to good use on all manner of traffic over second-class routes.

Once again an L.M.S. Standard boiler was chosen by Riddles and his team as the basis of the design, in this instance the boiler fitted to Fairburn's suburban-passenger 2-6-4T of the mid-1940s – a derivative of Stanier's earlier 2-6-4T design – suitably lengthened and kitted-out with B.R. auxiliary fittings. The short wheelbase B.R.2A-type tender was adopted generally for the Class 4, though this was displaced by larger alternatives where particular running conditions prevailed.

Erected at the celebrated ex-G.W.R. workshops at Swindon and numbered in the 75000 sequence, eighty Class 4 4-6-0s took to the rails from 1951 to 1957; an order for a supplementary ten engines for assignment to Eastern Region was abrogated owing to impending take-over of non-express passenger duties by diesel motor units. With smaller-diameter driving wheels and taller boiler mountings, the perky Class 4s had a daintier look about them than the squat, burly Class 5s, and their ultra-free steaming ability plus a willing readiness to meet any demands placed upon them endeared the engines to most of their operators. Whilst the single-pipe draughting arrangement common to the vast majority of B.R. Standard locomotives in the interest of simplicity was as effective in this application as in other classes, Western Region engineers experimented with dual blast-pipes and chimneys in a successful bid to increase performance, and several W.R.-based Class 4s were thus modified. In time this initiative was made official and selected Class 4 locomotives operating in other Regions were duly modified, including the entire Southern Region holding. These were fitted with a neater double chimney than the somewhat bulbous W.R. affair.

Lined black livery was the norm for Class 4s, while the precedent set by the Class 5s was imitated via appearance of several examples, in regular passenger service, in fully-lined Brunswick green. This colour scheme was particularly suited to the stylish medium-weight B.R. 4-6-0s.

Withdrawals commenced sporadically in 1964, the pace quickening in the ensuing four years, with the last few surviving Class 4s pensioned off in sight of the end of steam working in summer 1968. No less than four were transferred to enthusiast organisations, (including a W.R. dual-chimney version in express livery) an unusually high proportion in terms of numbers built, testimony to the outstanding reputation the 4-6-0s enjoyed in their heyday.

No. 75030 in unusually well-scrubbed state at Eastleigh shed yard, Southern Region, in September 1965, in concert with BR2A tender and attendant draught screens.
(F. Hornby)

CLASS 4 4-6-0, 1951

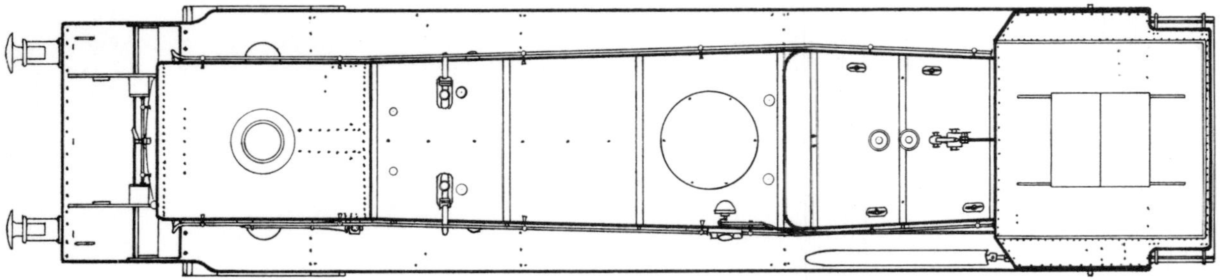

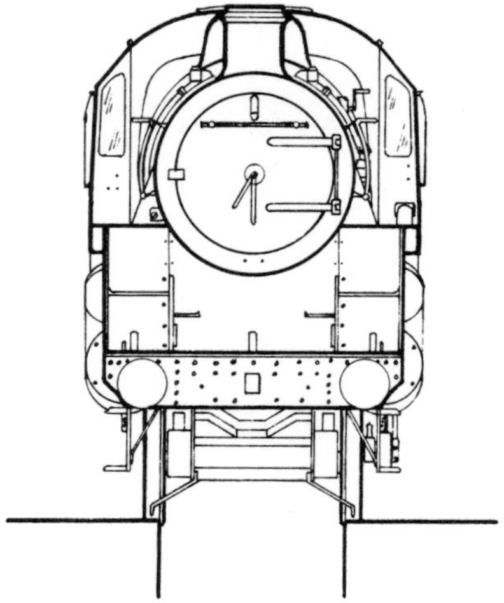

22

CLASS 4 4-6-0, 1951

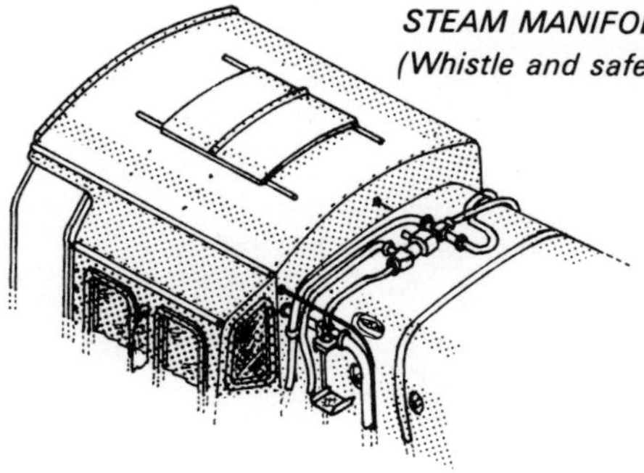

STEAM MANIFOLD AND PLUMBING
(Whistle and safety valves omitted for clarity)

CLASS 4 4-6-0, 1951

CLASS 4 4-6-0 DOUBLE CHIMNEYS
Western Region version: massive size, designed to limit of loading gauge

Southern Region version: lower and smaller, eventually fitted to all S.R.-allocated Class 4 4-6-0s

FRAME DETAIL

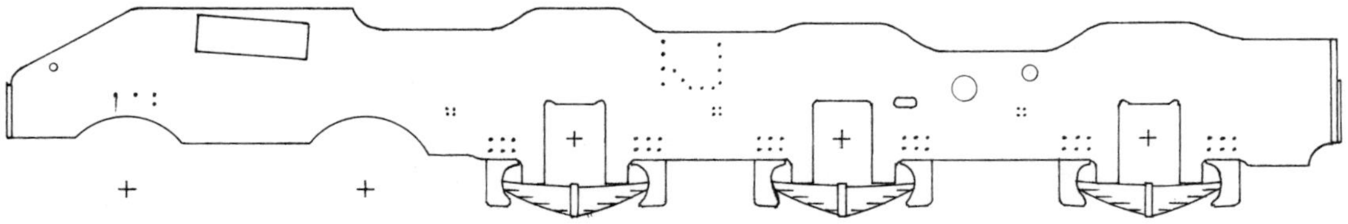

Quite easily mistaken for the L.M.S. 2-6-4Ts from which they were evolved, the B.R. 2-6-4Ts were the tank counterparts of and possessed many parts interchangeable with the Class 4 4-6-0s. Intended for suburban passenger and limited-distance goods duties, the big tanks were as popular as their tender cousins regarding their exemplary controllability and free-running characteristics. The stop-start nature of suburban passenger traffic demanded substantial accelerative powers and commensurate speed potential, especially crucial on densely-worked commuter lines where timing between trains was, and still is, a critical factor in efficient operation. The ability of a locomotive to run – and moreover be driven – equally effectively in either direction and therefore obviating turning at the end of a journey favoured the self-contained tank style, a concept recognised and implemented from the outset of railway history. This came to ultimate fruition in the very stable 2-6-4 wheel arrangement (the rear bogie permitting increased coal storage over the 2-6-2 type) and was employed on all Groups bar the G.W.R. The B.R. 2-6-4T was the conclusive expression of this philosophy and its utility was reflected in the protracted construction programme. Numbered in the 80000 series, 155 Class 4 tanks emerged between 1951 and 1956 from three erecting centres, Brighton (ex-S.R., formerly L.B.S.C.R.), Derby (ex-L.M.S., formerly M.R.) and Doncaster (ex-L.N.E.R., formerly G.N.R.) though superficially strange in view of the Brighton connection none of the output ever worked on the Southern Region, already well served by a long-established electrified commuter system.

Riddles' design featured the boiler from the L.M.S. Fairburn 2-6-4T adapted to B.R. specifications – the Fairburn model in its entirety had been proposed as a B.R. Standard but, apart from other considerations, it could not conform to the mandatory B.R. loading gauge limits. The principle of countrywide distribution inherent in B.R.

locomotive policy necessitated formulation of two specific loading gauges, labelled L1 and L2 covering every area of operation – and a unique distinction of the B.R. 2-6-4T was the curvature applied to tank, cab and bunker flanks to maximise space usage within the L1 gauge.

Although these 2-6-4 tanks gave a magnificent account of themselves in their allotted sphere of activity, the introduction of electric and diesel-propelled suburban passenger trains from the late 1950s onwards caused dispersal of the fleet onto lesser duties. Compulsory retirement came between 1962 and 1967, with five actively preserved for posterity.

The purposeful personality of the massive 2-6-4Ts depicted by Brighton-built, 1956-vintage No. 80137, nicely on the boil at Neasden, Eastern Region, in March 1957. *(F. Hornby)*

CLASS 4 2-6-4T, 1951

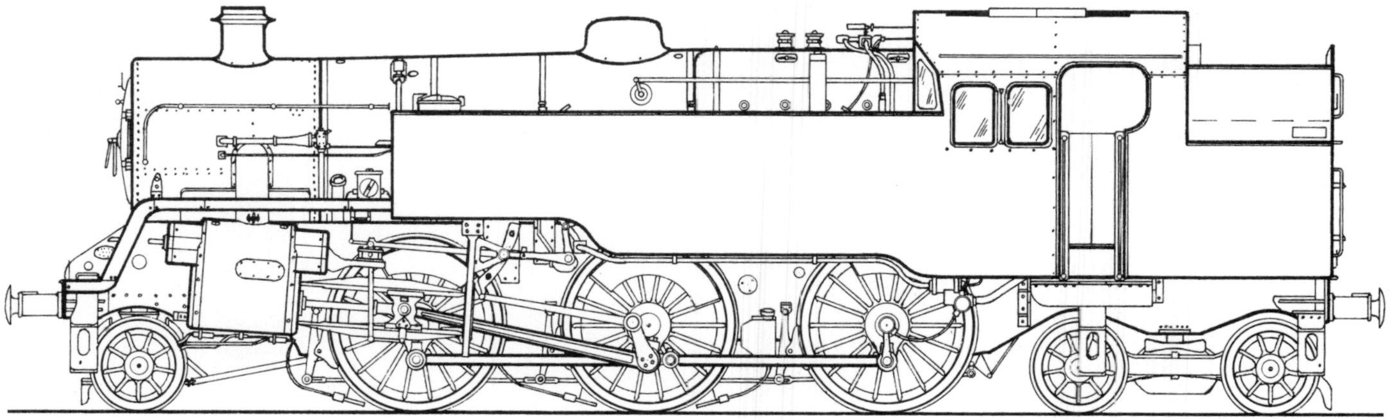

26

CLASS 4 2-6-4T, 1951

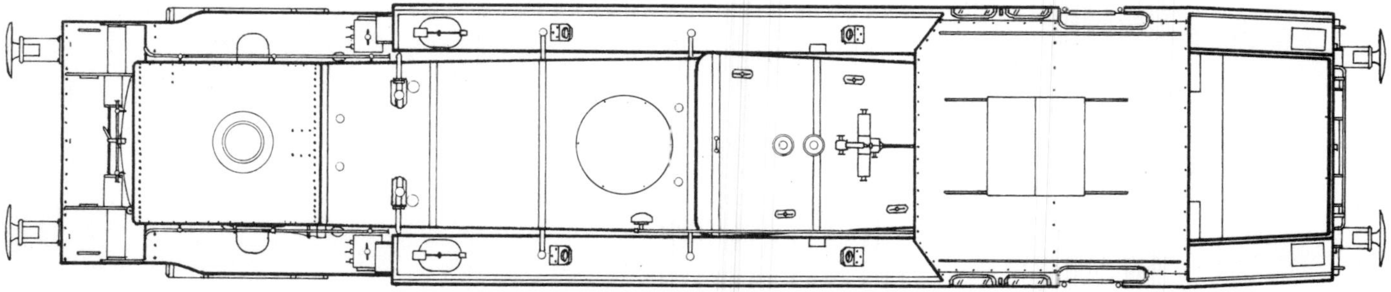

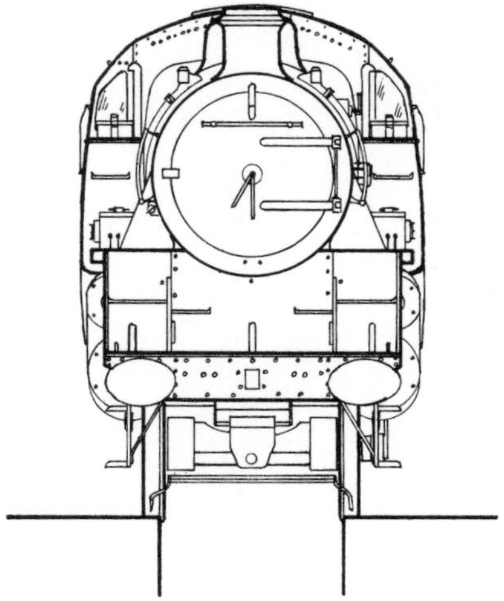

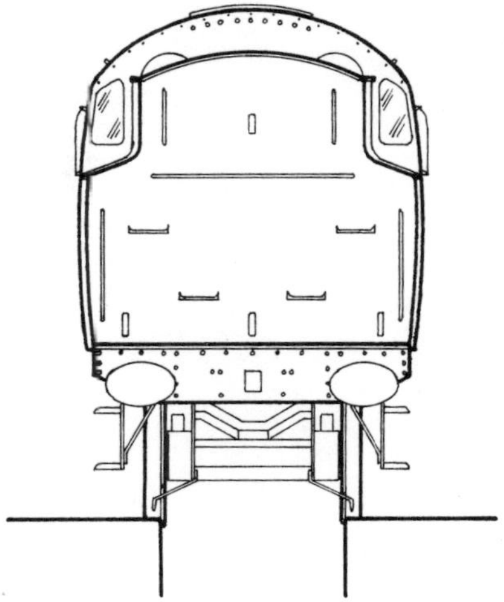

CLASS 4 2-6-4T, 1951

FRAME DETAIL

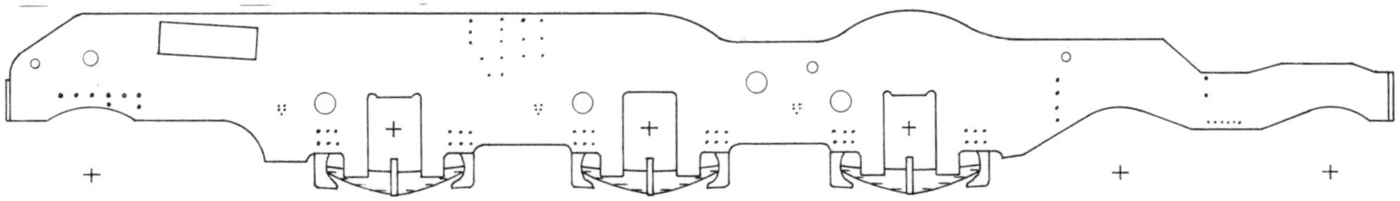

CLASS 4 2-6-0, 1952

Rather neglected in railway enthusiast circles compared to the similarly classified 4-6-0s and 2-6-4Ts, the 2-6-0 design was a direct descendant of Ivatt's very successful L.M.S. Class 4 2-6-0, production of which had started within weeks of Nationalisation and terminated shortly before unveiling of the B.R. version in late 1952.

The exigous, American-influenced profile of the Ivatt design was noticeably softened by alterations to running plate, cab and tender styling in the B.R. pattern, and indeed these adaptations wrought a very considerable transformation to the overall appearance. Cylinders, valve linkage and wheels were identical to those of the precursive 2-6-0s. Unlike the latter, double blast-pipes and chimneys were neither warranted nor fitted to the B.R. Moguls, and in any event the initial Ivatt locomotives so fitted had been modified into the simpler format at an early stage in their career.

The lighter weight of the Moguls relative to the Class 4 4-6-0s contributed to more comprehensive running powers; however, the reduced tractive effort resulting from less capacious boiler and cylinder dimensions, albeit partially offset by coupled wheels 5 inches smaller in diameter, ensured their use on humbler duties. Nonetheless the eager 2-6-0s proved as adroit on cross-country passenger haulage as their more potent confreres.

The versatility of the engines was confirmed by a production total of 115 examples outshopped from two sources, Doncaster and Horwich between 1951 and 1957. Numbered in the 76000 sequence and allocated to all Regions bar the Western, the Moguls were coupled to type B.R.2A tenders, the structure of this latter was strongly influenced by Ivatt's design for the parent 2-6-0, excepting those locomotives entrusted to the Southern Region where the lack of water troughs prompted use of enhanced-capacity B.R.1B alternatives.

Long before economical and mechanical obsolescence overcame them, the B.R. 2-6-0s were phased out of service in 1964-67; two escaped the ignominy of the cutting torch and survive in sympathetic hands.

Doncaster-built 2-6-0 No.76060, trailing a BR1B tender, at rest among the weeds in Eastleigh shed yard, Southern Region, March 1960. *(F. Hornby)*

CLASS 4 2-6-0, 1952

CLASS 4 2-6-0, 1952

30

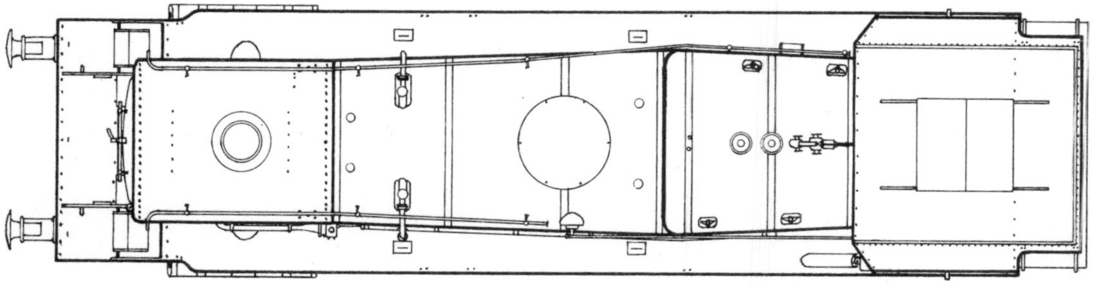

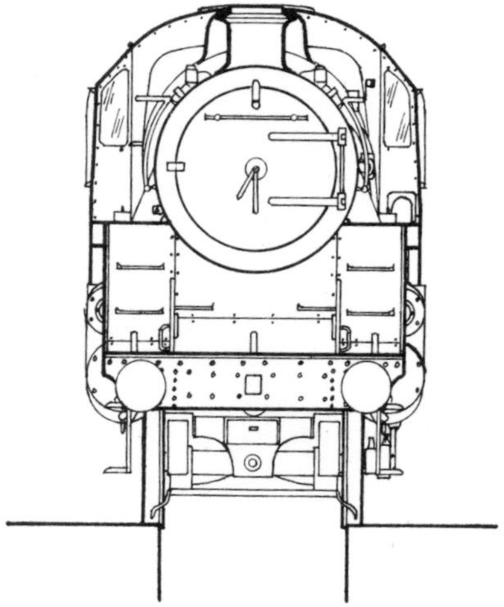

Bearing a family resemblance to the heftier Class 4 2-6-4Ts without the latter's curved side panels or angled front cab windows, the Class 3 2-6-2Ts were propagated for general-purpose operation with emphasis on passenger working, over routes closed to Class 4 locomotives and where Class 2 locomotives were insufficiently powerful for handling the more arduous duties.

Since there was no appropriate L.M.S. boiler to fall back on, Riddles opted for the excellent Swindon boiler utilised on former G.W.R. 5100, 6100 and 8100 2-6-2 tank classes, altered for B.R. installation by fitment of a dome, enlarged superheater, Southern Railway-style clack valves, and safety valves repositioned atop the Belpaire firebox.

45 were built at Swindon between 1952 and 1955, numbered in the 82000 series, and initially distributed to Southern and Western Regions. Several were outfitted in both lined and unlined B.R. (ex-G.W.R.) green in view of their passenger traffic connections but were destined to enjoy an abruptly curtailed term of service in their intended sphere, due not only to inroads made by diesel motor units but also to upgrading of tracks over which the tanks presided, with subsequent introduction of Class 4s. An order for eighteen engines supplementary to the 1954 assembly schedule was cancelled accordingly. The 2-6-2Ts were thus deployed onto more mundane tasks and gradually dispersed to Regions beyond their original boundaries.

Very akin in external appearance to the smaller Class 2 2-6-2Ts but distinguishable – beyond a more massive boiler and lower boiler mountings – by large shields protecting the prominent smokebox-flanking lubricators and boiler-top handrail added later, forward of the dome, the Class 3 tanks were scrapped in 1964-67 without a representative preservation.

In pristine lined passenger green livery, and cylinder cocks ajar, No.82039 stands outside Swindon workshops, following overhaul in November 1957. *(F. Hornby)*

CLASS 3 2-6-2T, 1952

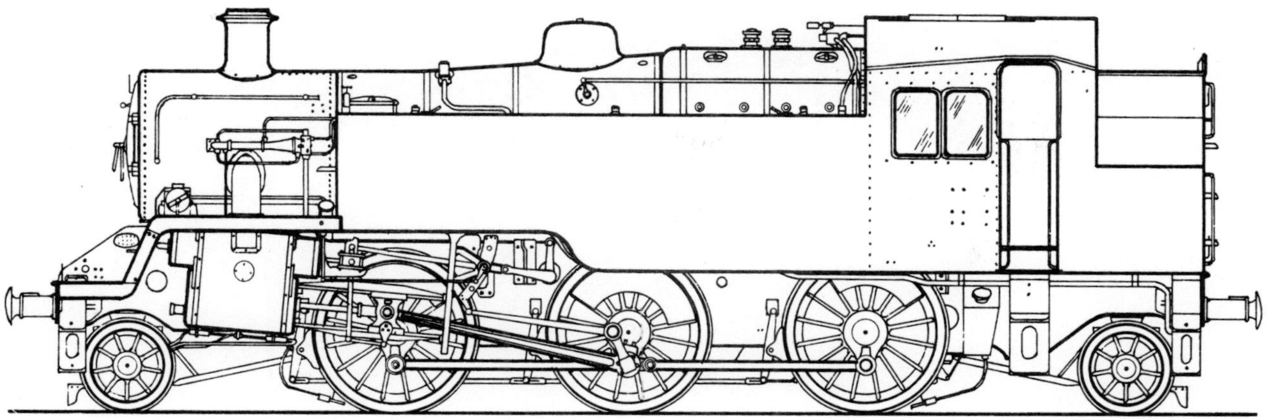

CLASS 3 2-6-2T, 1952

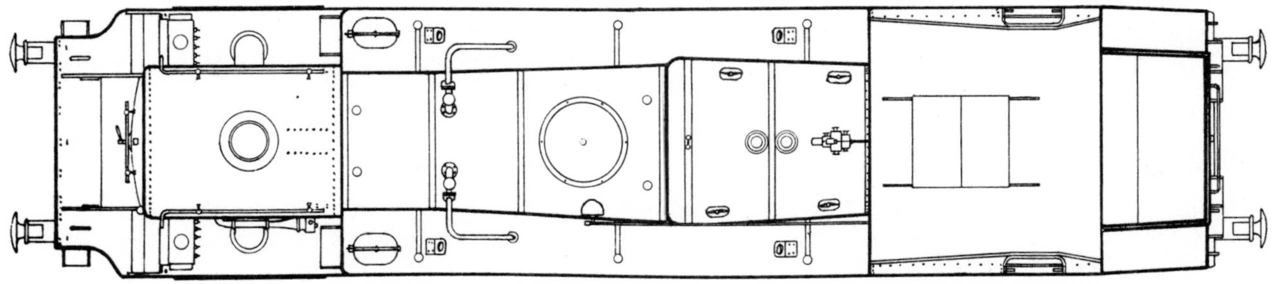

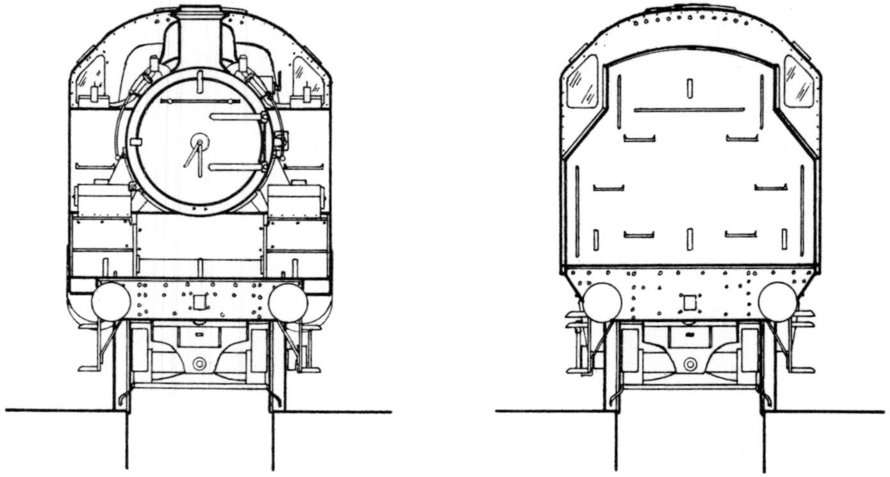

CLASS 3 2-6-2T, 1952

FRAME DETAIL, LIVE STEAM INJECTOR; DOME AND FIREBOX BACKPLATE EXPOSED

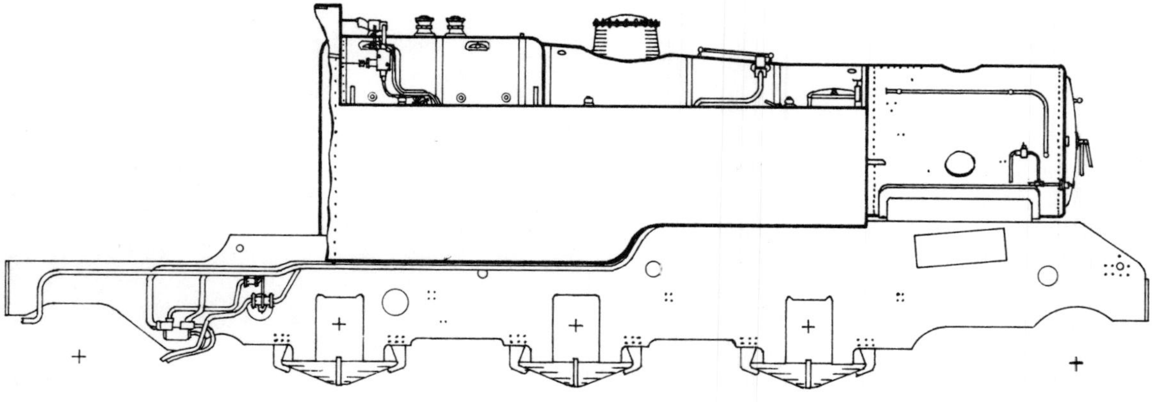

Found on all Regions excepting the Southern and the smallest in every respect of B.R. Standard tender locomotive classes, the Class 2 2-6-0 was an even closer derivation of Ivatt's L.M.S. precursor than the B.R. Class 4 2-6-0 was to the corresponding L.M.S. engine. Indeed only nominal modifications were made to bring the 1946-vintage design into alignment with B.R. design policy and the end result was all but indistinguishable.

128 examples of the Ivatt 2-6-0 had been turned out from 1946 until 1953, overlapping production of the B.R. counterpart for a short while, and the yeoman service they delivered on branch lines and general light traffic haulage justified extension of the fleet with updated models. Whereas draughting modifications gave the B.R. versions fractionally greater power, the high degree of reliability and overall mobility evinced by their L.M.S. forebears was consistently maintained.

Beyond the familiar B.R. boiler fittings, the external points of recognition on the B.R. Class 2 Mogul vis-à-vis its L.M.S. counterpart were the presence of a transverse fillet beneath the smokebox connecting side and front running plates; a cab with raised guttering, smaller windows, canted upper side panels and straight-edged cutaways beneath the running plate level (as opposed to radiused front corners); and a tender – codified B.R.3 – identical except for roller bearing axleboxes and minimal reprofiling of the cab structure to match that of the locomotive. The snug tender cab, complete with rearward-facing windows, greatly facilitated crew comfort on tender-first running, an important feature of branch-line operation where turntables rarely abounded.

No.78000, the introductory locomotive of the 65 in the class constructed at Darlington from 1952 until 1956, was repainted in unlined Brunswick green in lieu of the conventional black following major repairs at Swindon in the late 1950s, the sole exception in this case. Unlike the two Class 3 locomotive types, the fundamental motivation for construction and use of the Class 2s was never compromised, until displacement by diesel equivalents under the modernisation scheme spelt the extinction of branch-line steam power: the B.R. Class 2 Moguls were committed to the scrapheap between 1963 and 1967. A single specimen continues to operate preserved steam trains, in company with four L.M.S. classmates similarly rescued from dereliction.

Abandoned at the end of its career in an outlying yard at Toton, L.M. Region, in October 1966, grubby No.78055 forlornly awaits transference to the scrapyard.
(F. Hornby)

CLASS 2 2-6-0, 1952

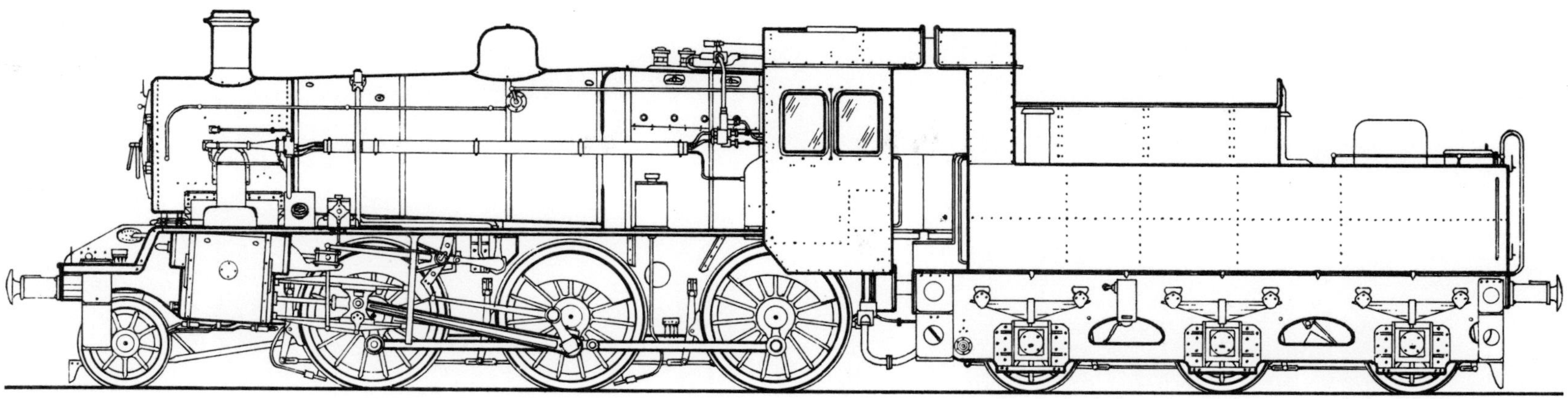

CLASS 2 2-6-0, 1952

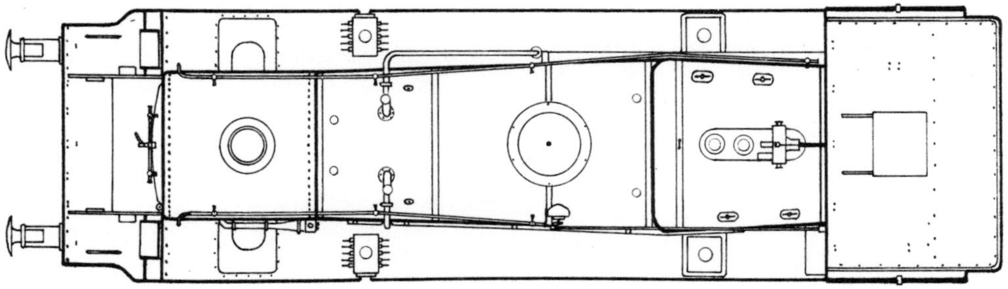

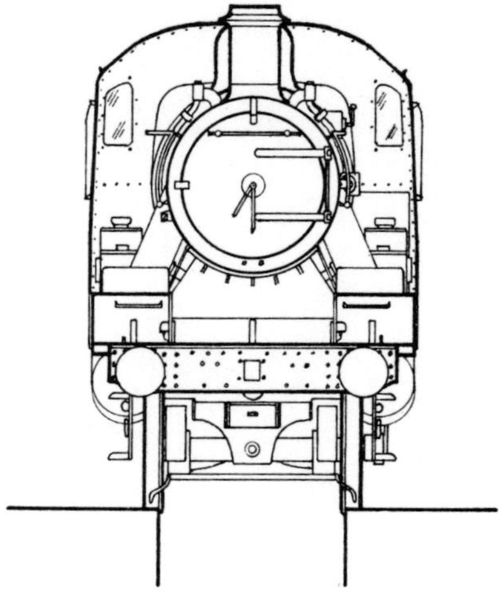

CLASS 2 2-6-0, 1952

FRAME DETAIL, FIREBOX BACKPLATE LOCATION

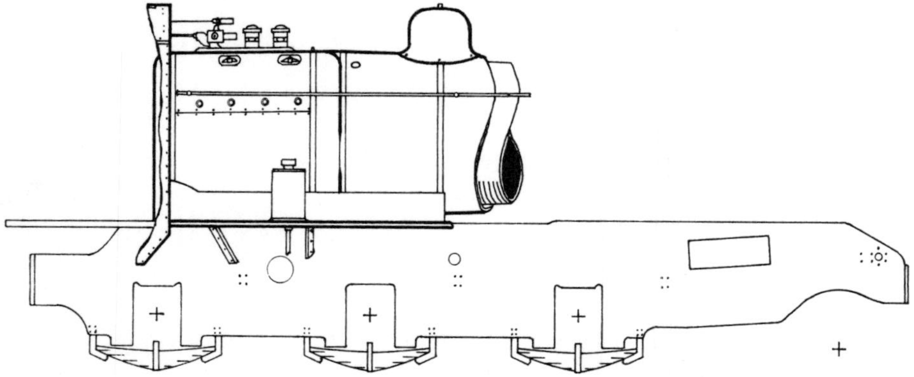

In concert with his lightweight Moguls, L.M.S. Chief Mechanical Engineer H.G. Ivatt had designed a Class 2 tank variation for push-pull branch passenger working, the two locomotives being simultaneously out-shopped from 1946 into the early 1950s. In similar fashion to the 2-6-0, the 2-6-2T was adopted as a B.R. Standard with minor specification changes and assembly of twenty B.R. Class 2 tank engines took place at Crewe in 1953 (output of L.M.S. 2-6-2Ts having ceased in 1952). A further ten locomotives appeared in 1957 from Darlington, their construction having been delayed four years owing to pressure of more urgently needed classes. However, their utility was undermined by diesel-powered alternatives almost from the moment they left the old-established former N.E.R. works. Given numbers in the 84000 series, merely thirty B.R. 2-6-2Ts filled a class already well populated by the Ivatt series; the earlier Crewe-built batch were assigned to London Midland Region and the Darlington products joined their progenitors on working various branches throughout the Southern Region.

Ivatt's principal reason for promoting his 2-6-2Ts in the first place was the presence on L.M.S. branch lines of archaic 0-4-4Ts in the old Midland Railway format, some of which dated back to the 1880s and thus were ripe for replacement. The Southern Region had likewise inherited sundry machines of similar antiquity and wheel configuration, hence allocation of relief L.M.S./B.R. 2-6-2Ts thereon. Their excellent steaming characteristics were a revelation to crews accustomed to asthmatic old stalwarts.

Twenty of the B.R. 2-6-2Ts were equipped for push-pull operation, the apparatus being removed on the locomotives' relegation to minor duties following dieselisation of branch services. Outward differences between the B.R. and L.M.S. Class 2 tanks were again confined to matters of detail; a fillet linking side front running plates, redesigned roof guttering, and footsteps with lengthy handrails in place of a ladder on the bunker back-plate constituted the changes, together with B.R. boiler mountings plus auxiliary fixtures. Withdrawal commenced in 1963 and terminated in 1965, a short existence indeed – especially so for the 1957 build. None has been acquired for preservation.

A Southern Region-allocated 2-6-2T, No.84025, straddles an inspection pit at Ashford shed in May 1958. Note the push-pull mechanism either side of the smokebox (the hooped object visible forward of the chimney is part of a water column.
(F. Hornby)

CLASS 2 2-6-2T, 1953

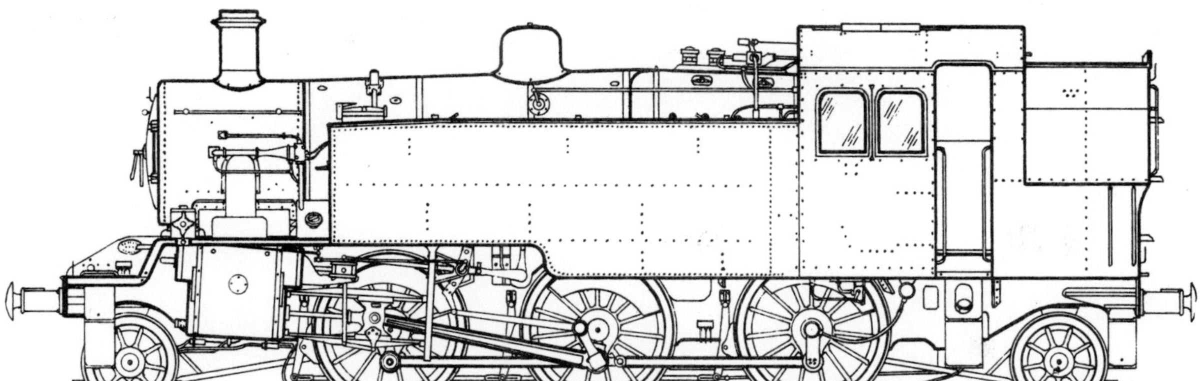

CLASS 2 2-6-2T, 1953

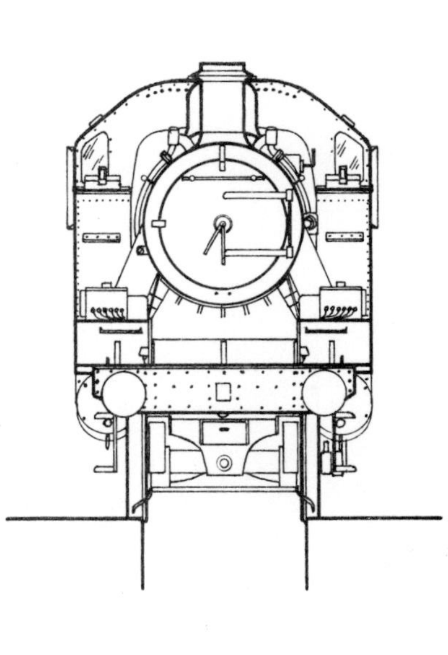

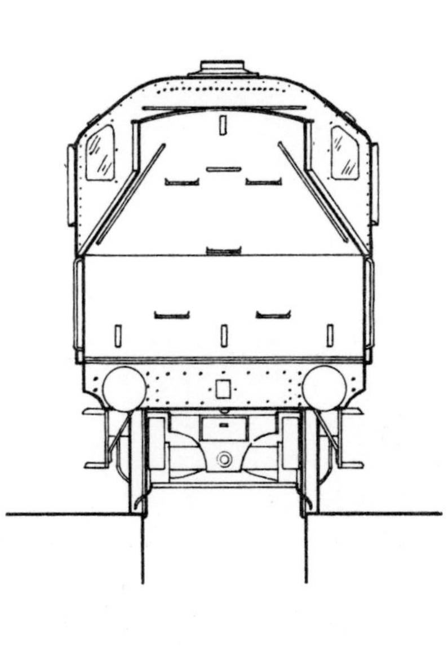

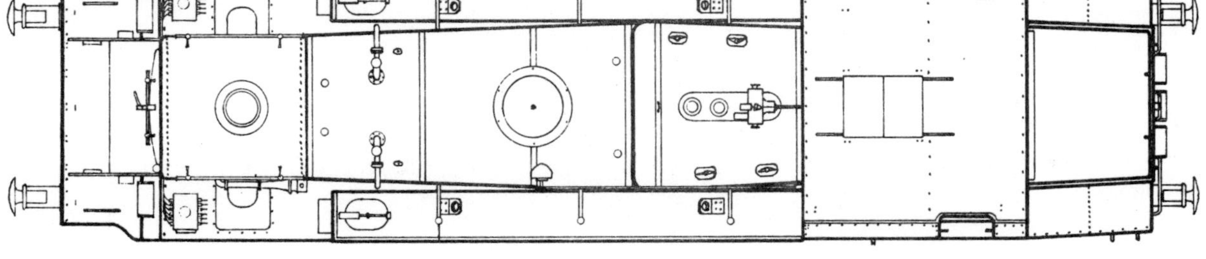

CLASS 3 2-6-0, 1954

The B.R. Standard locomotive class with the least number of constituents, excluding the one-off Class 8 Pacific, the Class 3 2-6-0 was the tender partner of the Class 3 2-6-2T. It was built around the selfsame G.W.R.-derived boiler, wheel sizes and coupled wheelbase, complete with type B.R.2A tender and modified cab/tender juxtaposition that eliminated the necessity for the rather clumsy anti-draught screens.

The total output of twenty Moguls, assembled at Swindon during 1954, were at first allotted to Eastern, North Eastern and Scottish Regions; their comparatively light axle loading permitted comprehensive route availability and, akin to the tank versions, the class was intended to handle minor-line traffic too weighty for Class 2 locomotives.

Supplementary orders were contemplated but, in the gestation period between commission and outshopping, events had overtaken the need for new engines in this power category and sphere of activity; improvements in track technology had permitted use of Class 4 machines on routes formerly the domain of less potent classes, thereby removing the basic raison-d'être of the 2-6-0s at the very outset of their career. Transferred in short order onto mundane duties such as shunting and pilot work, the Class 3s – numbered in the 77000 series – strayed beyond the areas of initial allocation until withdrawn between 1965 and 1967.

Outwardly very similar to the Class 4 2-6-0s and sharing the advanced steaming and running properties of most B.R. Standard designs, the untimely Class 3 Moguls may be said to have been the classic result of inadequate consultation between responsible B.R. departments.

Steaming gently outside Hamilton shed, Scottish Region, in June 1958, 2-6-0 No. 77006 stands ready for action.
(F. Hornby)

CLASS 3 2-6-0, 1954

44

CLASS 3 2-6-0, 1954

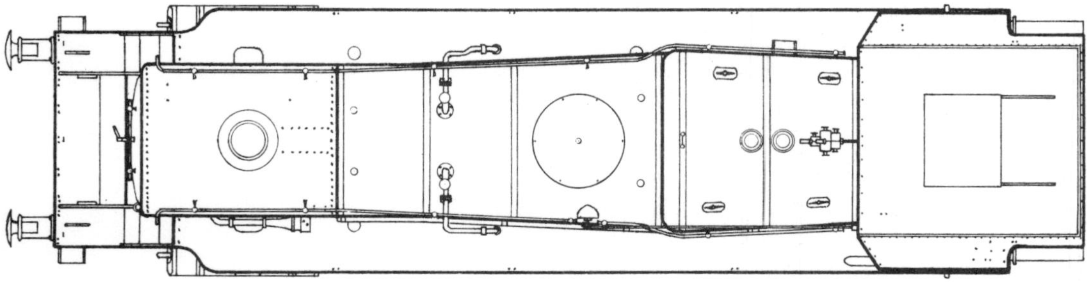

TYPE BR 2A 3,500-GALLON TENDER

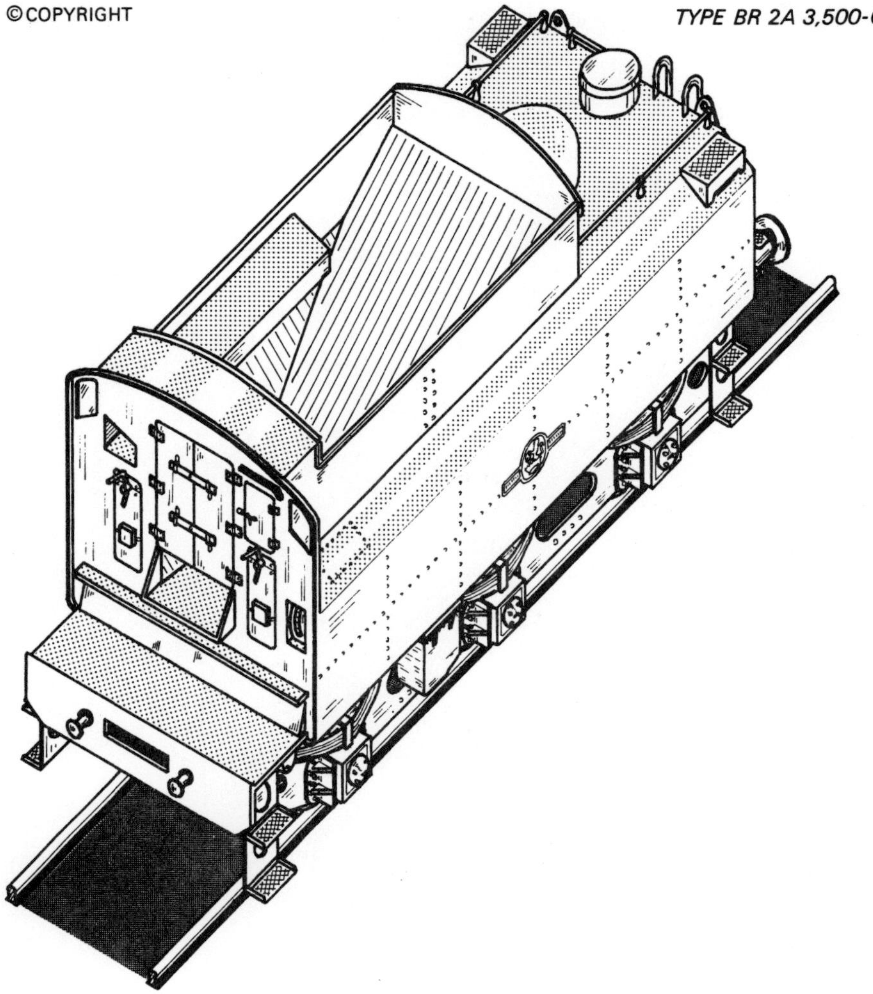

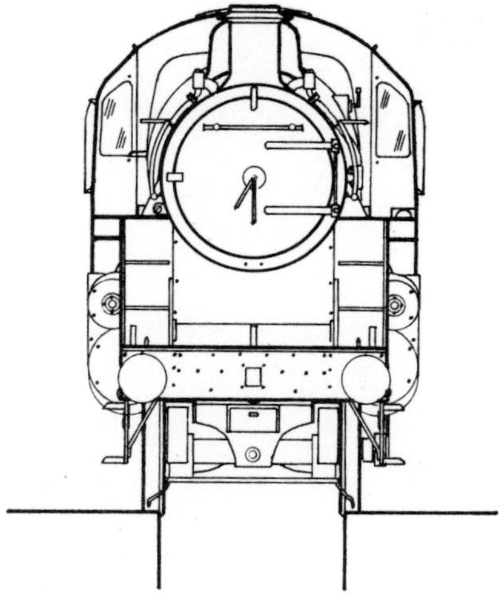

CLASS 3 2-6-0, 1954

FRAME DETAIL

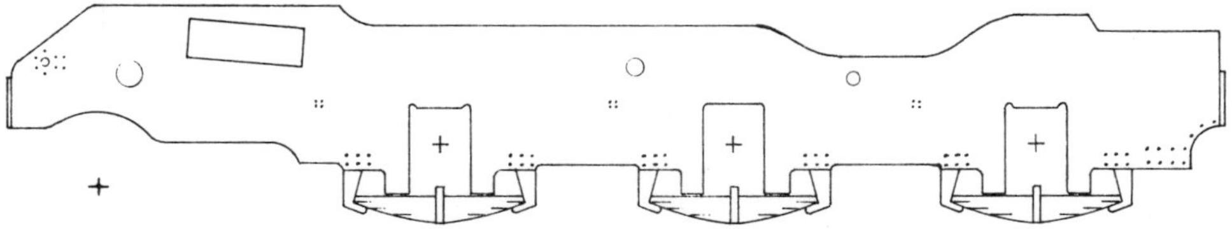

Another example of the predilection for survival of steam against all odds exhibited in high echelons of B.R. engineering management, the prototype B.R. Class 8 Pacific emerged at a stage when public announcement of the switch to diesel traction was imminent and thus the time and effort involved in its conception and construction, not to say development, was in the circumstances essentially wasteful. Nevertheless the story of B.R. Standard steam would be much the poorer had the design not been realised, for it heralded a promising trend in steam technology.

In the early 1950s the notion of a pure express passenger Pacific of greater power than the Britannias was strictly superfluous, since the countrywide roster of top-link locomotives was already adequate for all intents and purposes, only the holding of ex-L.M.S. four-cylinder 4-6-2s on the London Midland Region being under strength. With the benefit of hindsight, building of supplementary engines to the popular and well-proven L.M.S. design would have been the most expedient solution in the light of eventual dieselisation, yet this rather isolated Regional requirement provided the spur to sending Riddles' design panel onto a tangential course.

The resultant 4-6-2, No.71000 'Duke of Gloucester', erected at Crewe in 1954, was not only unique amongst its B.R. Standard stablemates in being the solitary example of its class; three cylinders, Caprotti rotary valve gear, dual blastpipes and chimney, and specially-conceived type B.R.1J tender with coal pusher, were also contributions towards non-conformity (whilst Caprotti motion and dual blasting were modifications made to one or other B.R. classes they were not original equipment thereon). In most other respects standardised parts were retained, indeed the locomotive was substantially derived from the Britannia format.

Having arrived on the scene so late in the day, the big Pacific suffered the fate of severely foreshortened development and total curtailment of a construction programme. Had the auguries been favourable, there is little doubt in view of its potential that the design would have met with complete success. As it was, numerous operational factors mitigated against any chance the locomotive stood of gaining acceptance over its tragically short working existence under B.R. ownership. Inevitably the deviant specification highlighted numerous defects that progressive development would have ironed out – principally a boiler (uprated from the Class 7 design) that consumed coal at a prodigious rate, combined with a steam output unequal to the demands of the cylinders – and the reaction to the newcomer among the crews accustomed to the superb Stanier Pacifics was less than appreciative. Nonetheless, in sympathetic hands, the locomotive proved very competent and some sparkling performances were generated on occasion.

Externally the Class 8 4-6-2 was similar to the Britannia, though several aspects marked it apart. The main departures from the progenitory layout were Caprotti valve gear (experience with which led to fitment on a late batch of Class 5 4-6-0s) – the outer valve chests heavily shielded – double chimney, tall-sided tender, cutaway front smokebox face, oval buffers, angled front footsteps and rear pony truck fitted with coil springs in an endeavour to combat the hard ride of laminated-sprung Britannias.

Named 'Duke of Gloucester' in honour of the then President of the International Railway Congress, the powerful Pacific was pensioned off in December 1962. Despite its tenuous place in the annals of B.R. traction, the locomotive was not scrapped forthwith but stored for several years more or less intact (minus the Caprotti mechanism which was eventually removed for museum display) at its home base of Crewe before transference to the renowned scrapyard at Barry – last resting place of so many British steam locomotives. Subsequently it was acquired by a Midlands-based preserved railway, where its restoration to full working order is in progress.

In store at Crewe works after withdrawal, 4-6-2 No.7100 stands neglected, though as yet complete, August 1965.
(F. Hornby)

CLASS 8 4-6-2, 1954

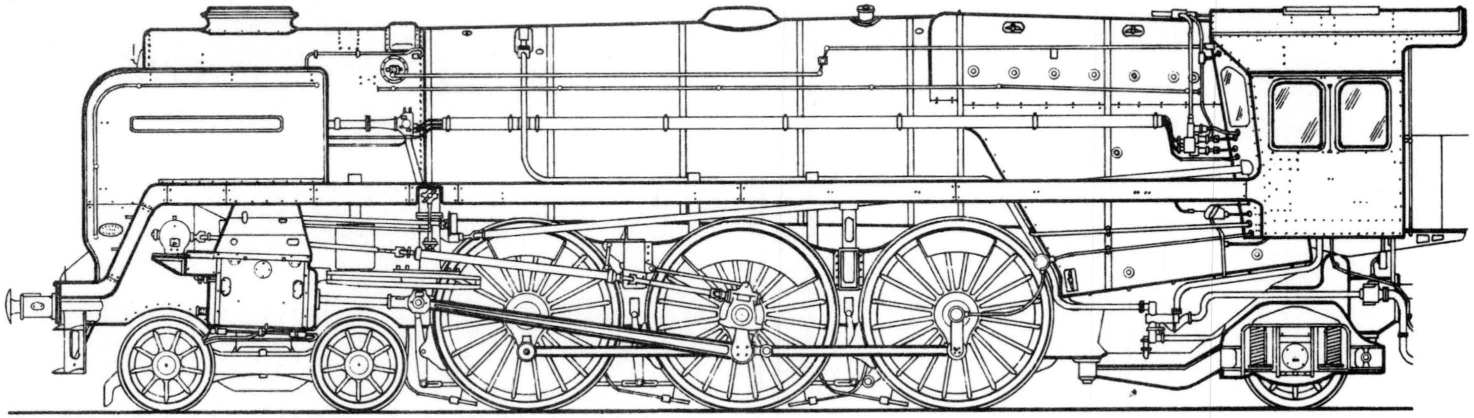

48

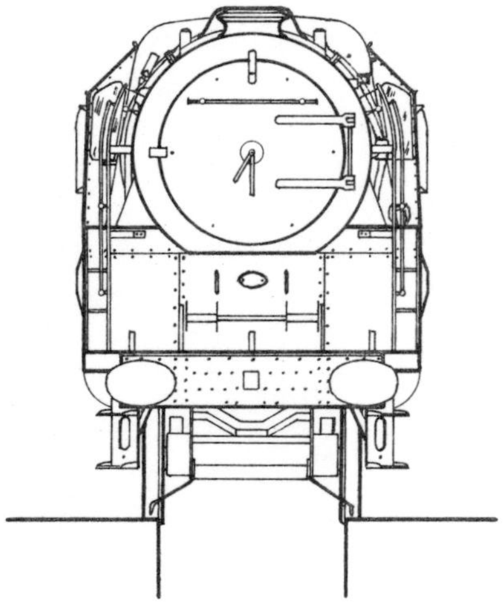

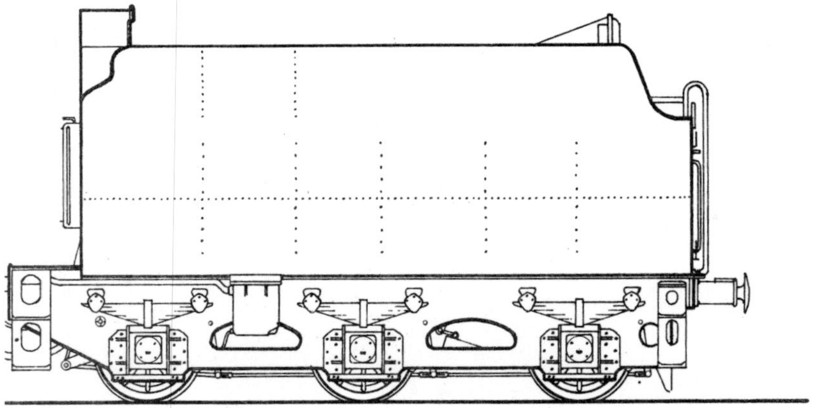

CLASS 8 4-6-2, 1954

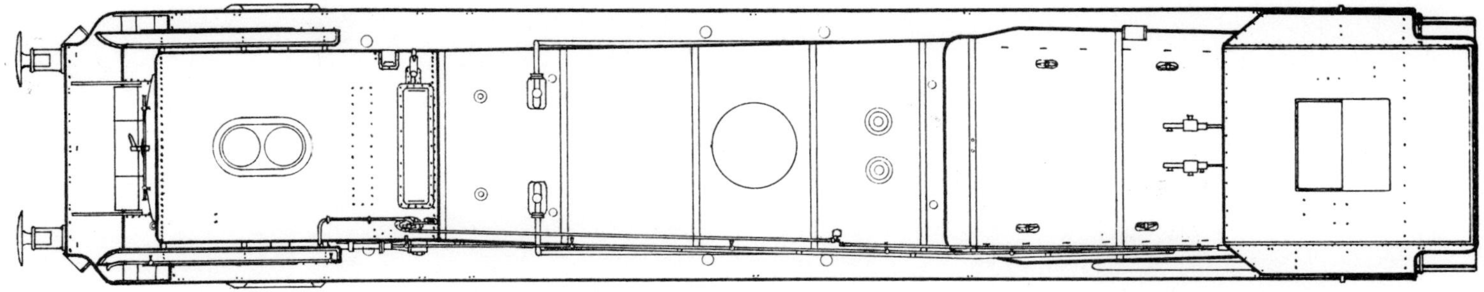

CLASS 8 4-6-2, 1954

RIGHT-HAND SIDE OF BOILER, PART-FRAME, TRUCK FRAME; AUXILIARY EQUIPMENT AND PIPEWORK SHOWN

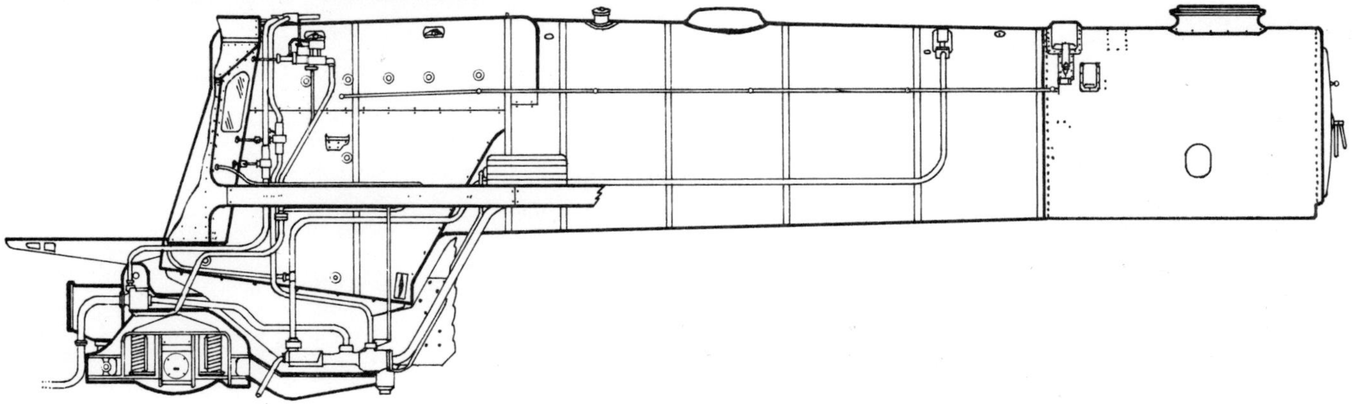

CLASS 8 4-6-2, 1954

CAPROTTI VALVE GEAR

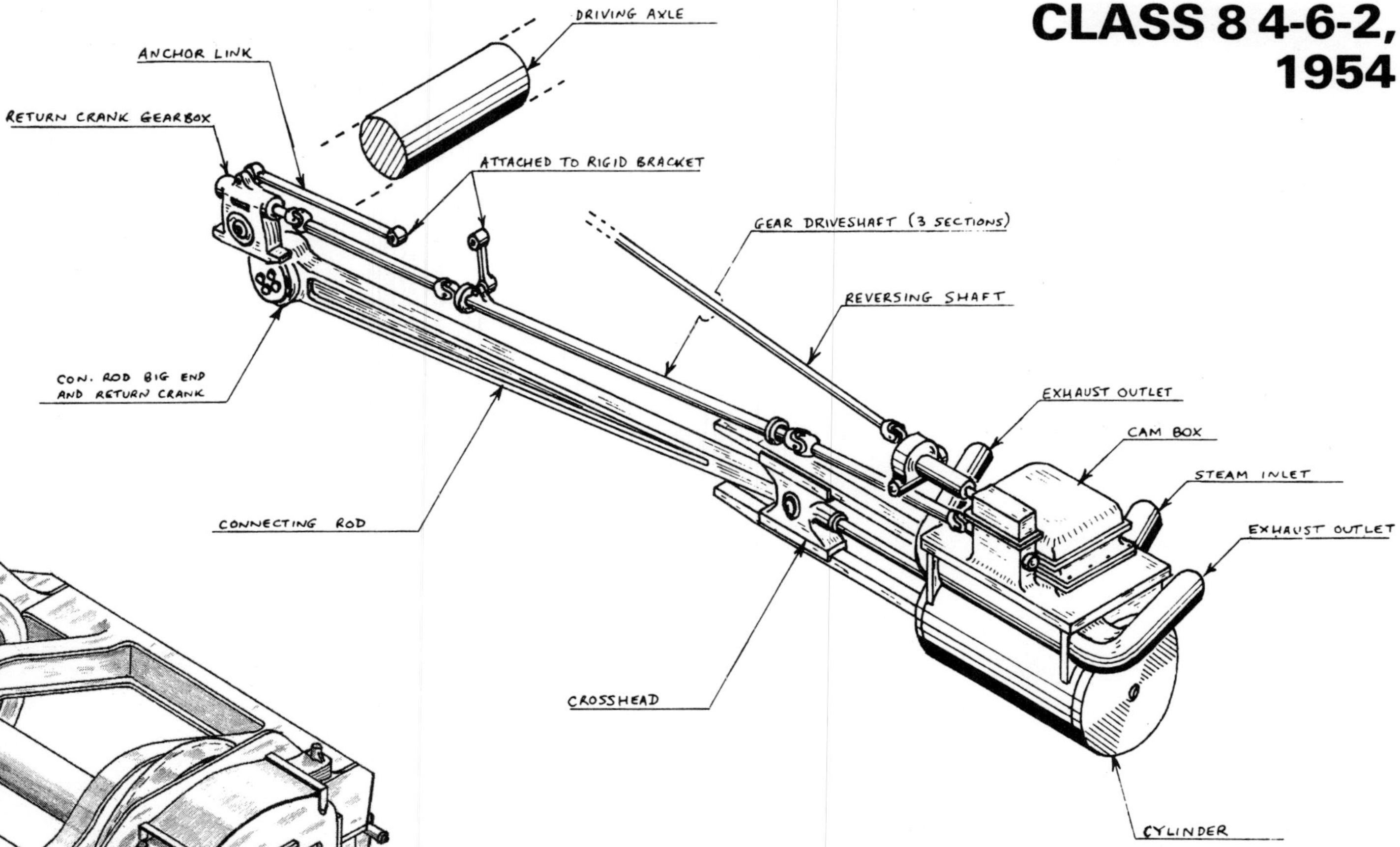

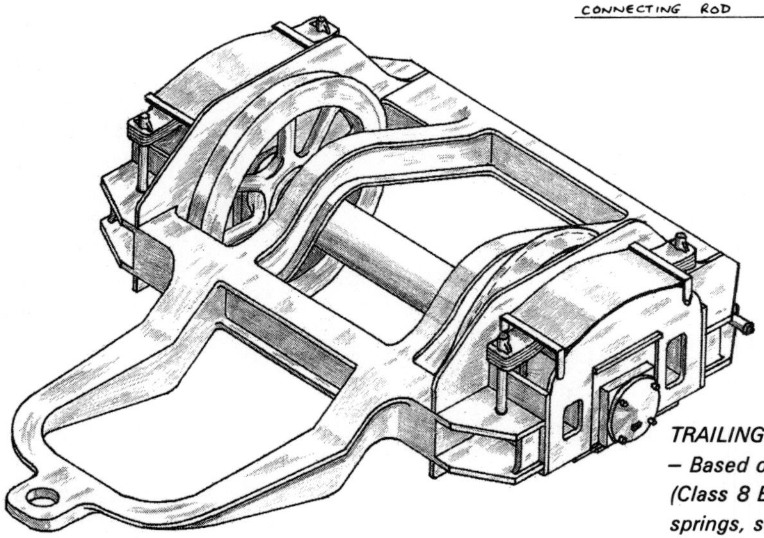

TRAILING TRUCK – B.R. STANDARD PACIFICS – Based on Bulleid's design for S.R. Pacifics (Class 8 B.R. 4-6-2 trailing truck suspended by coil springs, see p.49)

Arguably the best design to have emanated from Riddles and his colleagues, the Class 9 2-10-0 heavy freight design was the last component in the B.R. Standard steam collection.

In the planning stages for the freight locomotive Riddles had contemplated two possible wheel arrangements, 2-8-2 and 2-10-0. The disadvantages of the 2-8-2 style, though easier upon which to mount the desirable wide firebox, lay in the lower degree of adhesion compared with five coupled axles plus an ingrained prejudice stemming from Gresley's early, rather desultory L.N.E.R. 2-8-2s. The accomplishments of Riddles' wartime 2-10-0 design decided the issue. However there were two obstacles inherent in the layout; the boiler had to be unusually high-mounted for the firebox to clear the rearmost wheels, whilst the lengthy coupled wheelbase entailed use of flangeless centre wheels and ultra-high positioning of the cylinders on the frames so as to provide reasonable clearance with platform edges.

The boiler, with distinctive smaller-diameter smokebox and squat fittings, was individual to the class, whereas cylinders and draughting were based on the highly successful Britannia format; the result was an extremely powerful locomotive with supreme free-running characteristics and good controllability, assets essential for freight running where smooth progress was vital. Befitting the designated usage small (5-foot diameter) coupled wheels were installed for high traction in conjunction with theoretically low top speed; however, the Class 9s proved extraordinarily rapid for their specification and a maximum of 90 mph was unofficially recorded on passenger service. The lack of carriage-heating apparatus restricted the 2-10-0s to summer working on relief passenger trains, and a limit of 60 mph was rigorously enforced since excessive cylinder wear, the sole defect of consequence common to the class, was aggravated by the particular stresses imposed by passenger haulage. Eventually the 2-10-0s were banned from such working altogether.

Five tender types were employed, depending on Region and/or usage: B.R.1B, 1C, 1F, 1G and 1K. Of these the B.R.1K variety incorporated an automatic firebox stoking system intended to ease the heavy burden placed on firemen by the Class 9's high firing rate, but since the system required special coal, among other awkward particularities, it was abandoned after some years of trials.

Altogether a total of 251 2-10-0s were produced over six years from assembly lines at Crewe and Swindon, the ultimate member of the class being a landmark in railway history. Assembled at Swindon in 1960, it was the last steam locomotive built for use in Britain and was therefore honoured by an individual and highly appropriate name, selected from entries in a nationwide contest. Since the locomotive was also the last steamer out of what was probably the most famous of all railway workshops, a copper cap topped-off the double chimney and the magnificent Brunswick green express passenger livery was applied, in commemoration of the erstwhile Great Western Railway. Numbered 92220 – curiously, towards the end of construction numbers given to the engines were not necessarily assigned in chronological order, though all were within the 92000 sequence – 'Evening Star' was withdrawn after just five years active service. An obvious target for preservation, the locomotive is maintained in pristine condition by B.R. via the National Museum at York.

The sheer adaptability of the 2-10-0s in undertaking all manner of onerous tasks ensured a wide distribution throughout the Regions, although their weight restricted operation to main lines and upgraded secondary routes within the more commodious of the two B.R. loading gauges. The class as a whole quickly established an exalted reputation amongst locomotive crews. When the ex-Midland Railway 0-10-0 Lickey banker was laid to rest in 1956, irrevocably worn out after 37 years of exceedingly hard grind, a B.R. 2-10-0 took its place as principal banking engine on the notorious two-mile Incline, the unique-in-Britain smokebox-mounted electric headlight being transferred thereto.

69 2-10-0s were given improved draughting via dual blast-pipes and chimneys on building, and four single-chimney engines already in service were thus modified thereafter. The perceptible increase in performance shown by engines so equipped justified the extra expense involved, contrary to the experience gained with No.92250 fitted with a Giesl exhaust ejector in 1959 – though to be accurate, the advent of overall dieselisation was too close to warrant B.R. development of this Austrian device.

The large fleet of massive black 2-10-0s was phased out of commission between 1964 and 1968, the average lifespan being in the order of ten years, or approximately 20 per cent of that expected from a typical steam freight locomotive. Besides 'Evening Star', which holds such an important place in steam traction history, No.92203 has also been preserved in full working order, with further examples in process of purchase or restoration.

Exuding steam from every pore, 1955-built No.92050 (with BR1C tender) at Willesden shed, London Midland Region, in December 1958 amply illustrates the powerful bulk of its genre. *(F. Hornby)*

CLASS 9 2-10-0, 1954

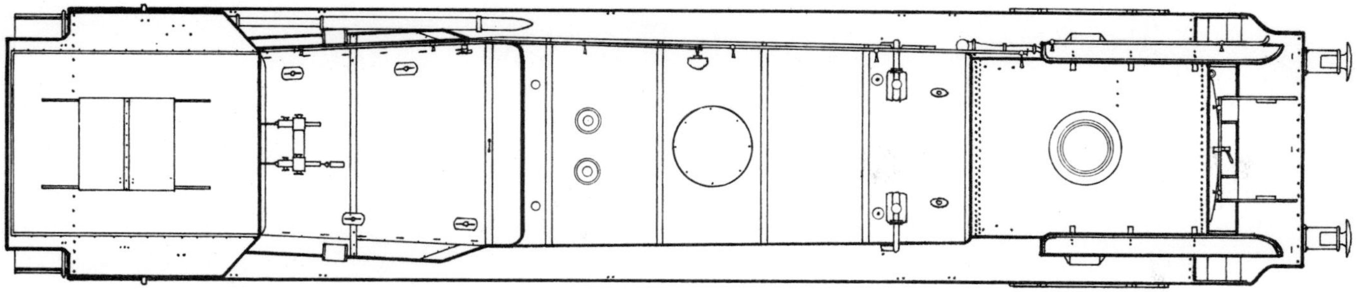

CLASS 9 2-10-0, 1954

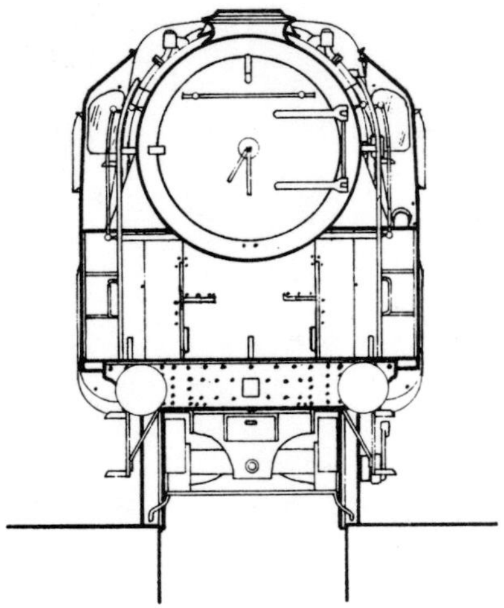

No. 92250
Giesl ejector chimney. Note also enlarged footstep beneath smokebox face, eventually fitted to most Class 9s

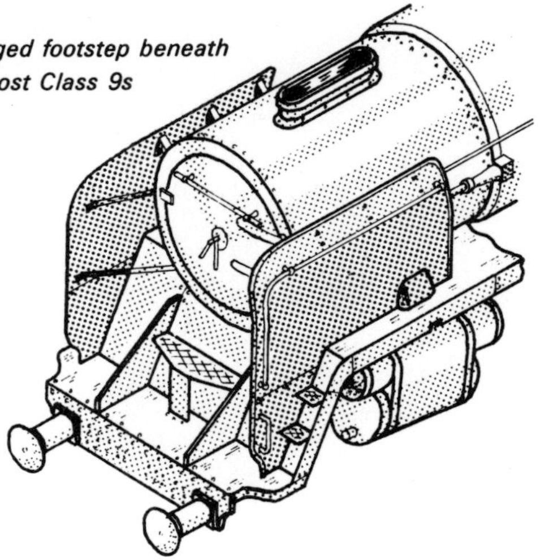

54

CLASS 9 2-10-0, 1954

FRAME, CUT AWAY TO REVEAL ASHPAN, AND BOILER/FIREBOX DETAILS, PLUS DOUBLE CHIMNEY

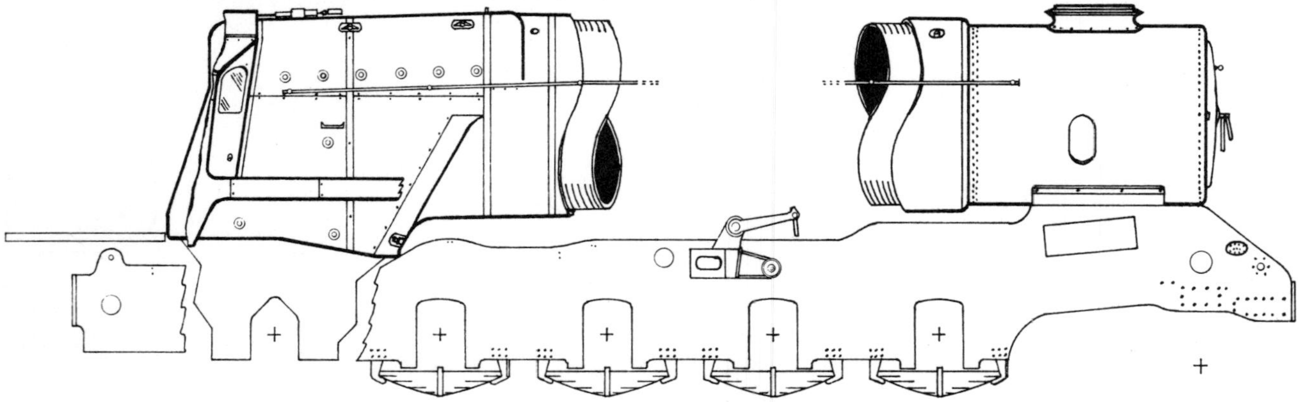

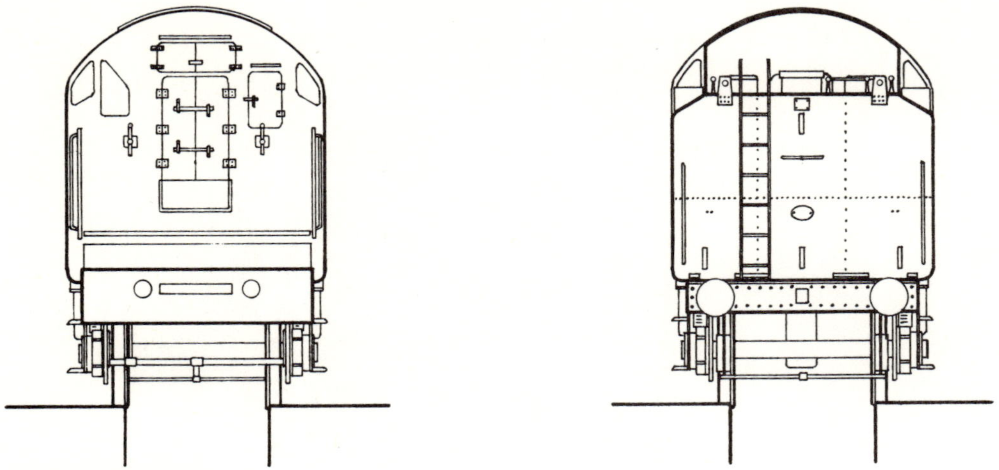

TYPE BR1G TENDER

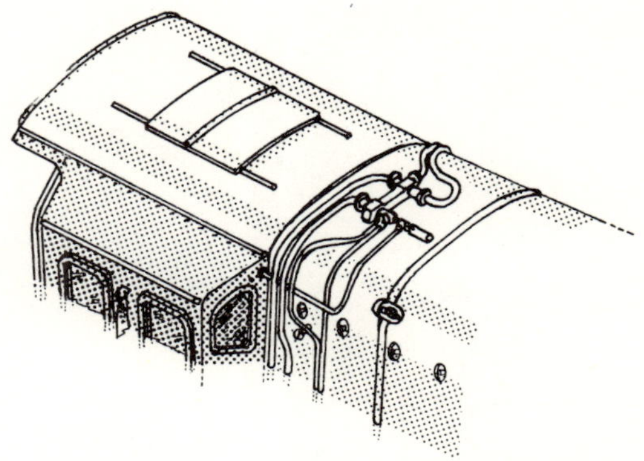

FIREBOX TOP PLUMBING AND STEAM MANIFOLD
(Basically similar on all B.R. Standard Classes)
Class 9 2-10-0

TYPE BR 1G
(For full side view, see p.53)
Coal capacity: 7 tons
Water capacity: 5,000 gallons

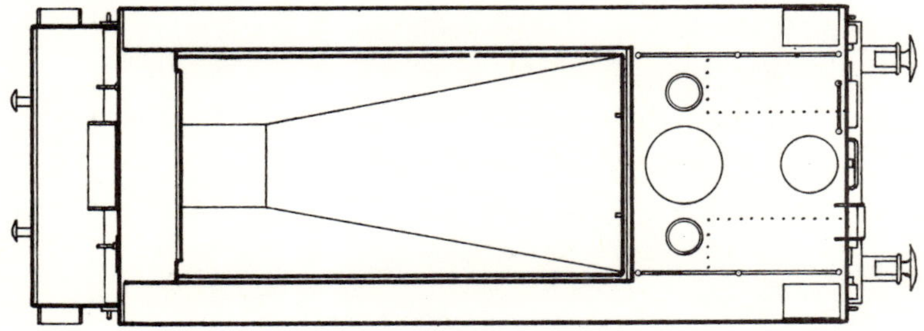

CENTRAL SECTION THROUGH TYPE
BR1G TENDER
(water scoop in operating mode)

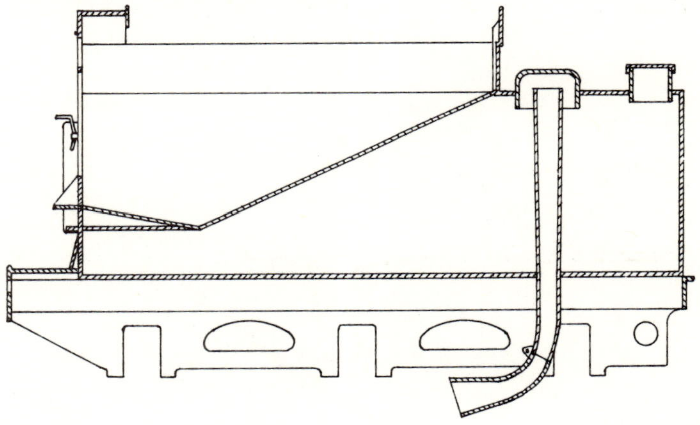

FRANCO-CROSTI BOILERED CLASS 9 2-10-0, 1955

During the mid-1950s the railway authorities encountered serious difficulties in acquisition of suitable coal for locomotive use, a recurring problem in twentieth century railway operation. Thus they were receptive to overtures from a company claiming favourable consumption rates for a specialised boiler design, invented in Italy and widely used on that country's nationalised network, in conjunction with a system of payment that involved B.R. in relatively light capital expenditure.

A batch of ten Class 9 2-10-0s within the 1955 construction programme was chosen for fitment with Franco-Crosti boilers and the locomotives were duly outshopped from Crewe and numbered 92020-29. Basically the Crosti method of steam production employed a conventional boiler for which feed-water was preheated in a secondary boiler, the latter being supplied with heat from the main boiler and exhausted via a laterally-located chimney (the normal smokebox chimney was used only in the process of heating the main boiler from cold). In Class 9 application the secondary boiler, with associated smokebox and door, was slung beneath the main boiler, which was set to a record height for a British steam locomotive and for which boiler mountings were extremely shallow to clear the loading gauge; the 'running' chimney was positioned on the right-hand side of the engine and in time shrouded by a deflector plate, following trouble with lingering smoke hampering crew vision inside the cab. Sited either side of the boiler on conventionally boilered B.R. locomotives, the topfeed clack valves on the Crosti versions were both attached to the righthand side of the boiler. Because of its extraordinary height and the limitations of the loading gauge, the main boiler was dimensionally smaller than the orthodox counterpart. Overall external appearance, especially when viewed from the 'chimney' side, was wholly unlike any other British steam locomotive and generally acknowledged as on a par with Bulleid's S.R. 'Q1' class for ugliness.

Unfortunately, for both parties concerned, the experiment was an outright failure. The expectations raised by its proponents were far from forthcoming in practice and within three years of construction all ten Crosti-boilered 2-10-0s had been relegated to storage in shed yards pending conversion. The locomotives were re-commissioned in 1959 following institution of normal draughting arrangements and removal of preheating boilers; the Crosti-inspired main boilers were retained intact with the strange smokebox 'wings' (through which flue gases were previously transmitted to the secondary boilers) so that the reconstituted machines were effectively less potent than their classmates, though officially remaining in the same power category. Not surprisingly, none was earmarked for salvage.

Depicting the unusual accoutrements of a Crosti 2-10-0, No.92021 awaits conversion in store at Wellingborough, London Midland Region, in May 1959. Note the smoke deflector and BR1B tender, common to Crosti versions. *(F. Hornby)*

FRANCO-CROSTI BOILERED CLASS 9 2-10-0, 1955

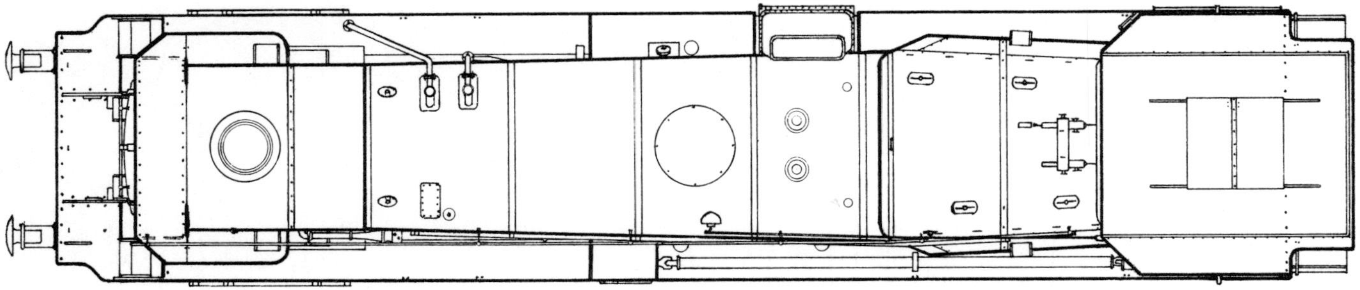

58 FRANCO-CROSTI BOILERED CLASS 9 2-10-0, 1955

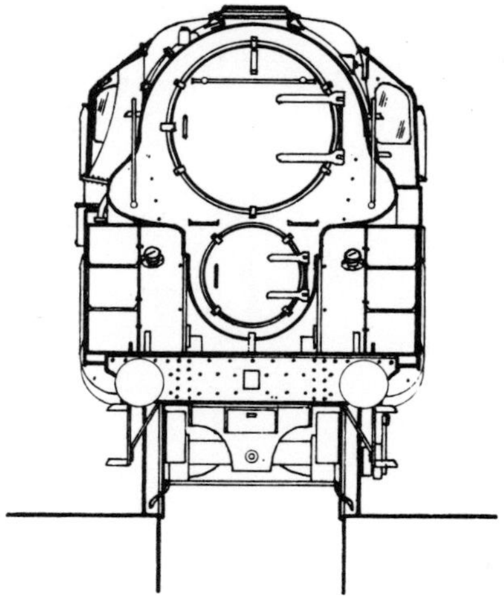

FRANCO-CROSTI BOILERED CLASS 9 2-10-0, 1955

RIGHT-HAND SIDE SHOWING PART-FRAME DETAIL, SIDE CHIMNEY, INJECTOR PLUMBING AND UNDERSLUNG SECONDARY BOILER OUTLINE

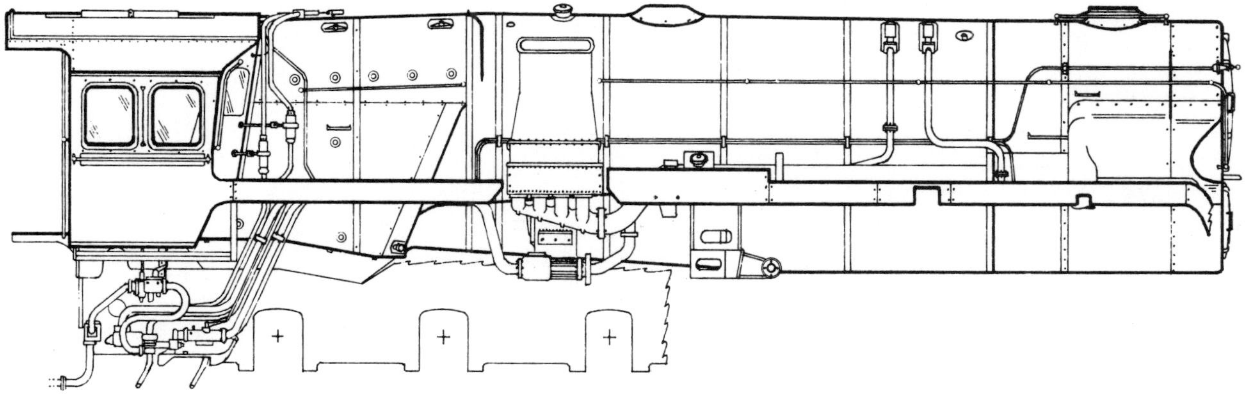

Southern Steam Doubleheaded Tony Fairclough & Alan Wills
Southern Steam in the West Country Tony Fairclough & Alan Wills
More Southern Steam in the West Country Tony Fairclough & Alan Wills
Southern Electrics edited by B.W. Rayner
Southern Branch Line Steam: 1 Tony Fairclough & Alan Wills
Southern Branch Line Steam: 2 Tony Fairclough & Alan Wills
Southern Branch Line Steam: 3 Tony Fairclough & Alan Wills
Southern Steam through the Years Tony Fairclough & Alan Wills
Steam on the Brighton Line edited by A.C. Perryman
Recollections of the Southern Between the Wars H.C. Casserley
Bodmin & Wadebridge Railway Tony Fairclough & Alan Wills
Southern Steam in Works Tony Fairclough & Alan Wills
Southern Steam in Close Up Tony Fairclough & Alan Wills
SR Working Timetable Appendices 1934

North Eastern Steam in Action Brian Morrison
North Eastern Steam from Lineside P.J. Lynch
North Eastern Steam in Northumbria Malcolm Dunnett
North Eastern Pacifics Norman E. Preedy
North Eastern Steam Locomotive Album P.J. Lynch
LNER Locomotives 1923-1948 H.C. Casserley
LNER Steam 1923-1948 H.C. Casserley

BR Diesels in Action edited by G. Weekes
BR Diesels in Action: 2 edited by G. Weekes
BR Diesels in Action: 3 edited by G. Weekes
BR Diesels in Action: 4 edited by G. Weekes
BR Diesels in Action: 5 edited by G. Weekes
BR Diesels on Shed N.E. Preedy
BR Diesels in Close-Up N.E. Preedy & H.L. Ford
BR Diesels in the Landscape Derek Cross
WR Diesel-Hydraulics Norman E. Preedy & G.F. Gillham
BR Diesels and Electrics around Britain A.W. Hobson
BR Diesels Doubleheaded edited by G. Weekes
Diesels in London and the Home Counties P.A. Dobson
Diesels on Eastern Region edited by Norman E. Preedy
Diesels on Western Region edited by H.L. Ford
Diesels on Scottish Region S. Rickard
Diesels on Midland Region edited by G. Weekes
Diesels in the Highlands edited by G. Weekes
Diesels in the West Midlands and Central Wales
Diesels on Cornwall's Main Line H.L. Ford
Diesels on the Devon Main Line edited by H.L. Ford
Modern Rail Album A.W. Hobson
Branch Lines Round Britain in the Diesel Era J.A.M. Vaughan
DMUs Countrywide P.J.Fowler
The Westerns: BR Class 52 Diesel-Hydraulics H.L. Ford
The Warships: BR Class 42/43 Diesel-Hydraulics H.L. Ford
BR Diesel Miscellany: 1 edited by G. Weekes
BR Diesels in Detail M. Oakley & Brian Morrison

Diesels West G.F. Gillham
Diesel Super-Power around the World R.K. Evans

BR Standard Steam in Action Derek Cross
BR Standard 2-10-0 Class 9F edited by G. Weekes
BR Standard Steam in Close-up Tony Fairclough & Alan Wills
BR Standard Steam in Close Up: 2 Tony Fairclough & Alan Wills
BR Standard Britannia Pacifics edited by G. Weekes
Memories of Steam around Britain P.J. Lynch
Scottish Steam in the 1920s R.D. Stephen
Scottish Steam Miscellany R.D. Stephen
Scottish Branch Line Steam edited by Jack Kernahan
Scottish Main Line Steam edited by Jack Kernahan
Steam in the Western Highlands edited by Jack Kernahan
Last Decade of Scottish Steam Derek Cross
Steam on the Waverley Route R.H. Leslie
Steam on Britain's Miniature Railways Robin Butterell
Steam on the Somerset & Dorset G.A. Richardson
Steam around Britain in the Seventies L.A. Nixon
Cornwall's Railways A. Fairclough
The Railways of Devon A. Kingdom
Specials in Action A.R. Butcher
Mineral Railways of the West Country Tony Fairclough
Industrial Steam Adrian J. Booth
British Steam around the Regions K.R. Pirt & D.E. Penney
The Tralee & Dingle Railway David G. Rowlands
European Narrow Gauge Steam: 1 D. Trevor Rowe
European Narrow Gauge Steam: 2 D. Trevor Rowe
Industrial Narrow Gauge Railways in Britain P.D. Nicholson
British Narrow Gauge Steam edited by Michael J. Messenger
More British Narrow Gauge Steam Michael J. Messenger
Steam on the SNCF Peter F. Winding
European Steam in Action P.J. Lynch
Austrian Steam from Lineside V.C.K. Allen
Portuguese Steam from Lineside V.C.K. Allen
Railways of the Republic of Ireland 1925-1975 M.H.C. Baker
Articulated Locomotives of the World Donald Binns
South American Steam M.H.J. Finch
Steam in India Hugh Hughes
Steam in the Andes Brian Fawcett
New Zealand Steam Spectacular Derek Cross
British-Built Steam Locomotives Overseas J.N. Westwood
Vulcan Foundry Locomotives 1932-1956 D.E. Gudgin
Electric Locomotives of the World F.J.G. Haut
Steam Super-Power around the World R.K. Evans
Light Railways around Britain H.C. Casserley
The later years of Metropolitan Steam H.C. Casserley
Irish Railways in the Heyday of Steam H.C. Casserley
The Midland Railway edited by The Midland Railway Trust
Steam in Australia C.B. Gilbertson
The Great Central Recalled George Dow
North British Steam Locomotives Overseas John H. Court
North British Steam Locomotives in Britain John H. Court

On the Welsh Narrow Gauge J.I.C. Boyd
Scottish Railways in the Heyday of Steam H.C. Casserley
European Trams L.F. Folkard
British Trams L.F. Folkard
Lynton and Barnstaple Railway J.R. Yeomans
Border Steam Peter Brock
Welsh Railways in the Heyday of Steam H.C. Casserley
On the Isle of Man Narrow Gauge J.I.C. Boyd
Narrow Gauge Steam out of Portmadoc – 25 years of the Festiniog Railway N.F. Gurley
Traction Engines in Close-Up E.H. Sawford
Municipal Buses in the 1960s L.W. Rowe
Midland Red Buses M.W. Greenwood

Paperbacks etc

BR Standard Freight Wagons D. Larkin
Private Owner Freight Wagons D. Larkin
Pre-Nationalisation Freight Wagons D. Larkin
BR General Parcels Rolling Stock D. Larkin
BR Departmental Rolling Stock D. Larkin

Class 44/46 Diesels (The Peaks) J.A. Walker
Brush-Sulzer Class 47 Diesel-electrics edited by B. Ringer
Electrics of West Coast Main Line Roly Longhurst
EE Class 50 Diesel-electrics Michael Oakley
BR-EE Class 37 Diesel-electrics D & E.G.
BR Diesel Shunting Locos D. & E.G.
BR Diesels Departed Derek Cross
BR Diesel Cavalcade Brian Morrison

BR Diesel Freight (Traditional Era) Michael Oakley
BR Diesel Freight (Era of Specialisation) Michael Oakley
BR 125 High Speed Train D. & E.G.
Named Locomotives of British Rail D. & E.G.
BR Electric Multiple Units Brian Hardy
Industrial Diesels – a pictorial survey A.J. Booth

LNER Footplate Memories Charles Meacher
When there was Steam – memo of a fireman Tony Barfield
On the Footplate – memo of a GWR Engineman J. Drayton
A Loco Fireman Looks Back: WR Recollections Ray Gwillam
Steam Supreme – recollections of Scottish Railways in the 1920s R.D. Stephen
Drawn by Steam – memories of a Tyseley locoman L.C. Jacks
Firing Days at Saltley T.R. Essery
Locomotives at War – Army railwav reminiscences of the Second World War P.M. Kalla-Bishop
Through the Links at Crewe – Top Link footplate memories: volume one 'Piccolo Pete' Johnson
St Philips Marsh – memories of an engine shed D.J. Fleming
Nothing Like Steam – footplate work on SR B. W. Aynsley
Living with Locos (LNER) Charlie Meacher
North Eastern Engineman – driver Syd Midgley and fifty years of steam P.W.B. Semmens
The Art of Railway Photographv – a pictorial manual of photo-technique L.A. Nixon

Cornwall's Railways A. Fairclough
Vale of Rheidol Pictorial J. Reeves
Welshpool & Llanfair Railway Pictorial Ralph Cartwright
The Cornwall Railway R.J. Woodfin
Spirit of Steam (New Zealand) Roy Sinclair
The London RT Bus P.J. Lynch
Redruth and Chasewater Railway D.B. Barton